PENGUIN BOOKS

Silent Spring

'The revolution in our attitude towards pollution and the extravagant use of chemicals to control pests in the countryside is largely due to just one individual – Rachel Carson ... She was something rare for those times – a scientist who was also a brilliant writer'
Lewis Wolpert, *Sunday Times*

'One of the first social critiques of modern industrial behaviour ... her tone and sharpness were luminous'
John Vidal, *Guardian*

'Her warnings against organophosphates seem as relevant as ever, in the wake of Gulf War Syndrome and sheep-dip poisoning ... she opened her readers' eyes to poetry and grandeur in the natural world, which they never realised to have existed' Clive Aslet, *Observer*

'A book which changed the way people looked at the world' Antony Rouse, *Spectator*

'She was the originator of ecological concern'
Doris Lessing, *Sunday Telegraph*

Rachel Carson (1907–64) wanted to be a writer for as long as she could remember. However, while at Pennsylvania College for Women (now Chatham College), she changed her major from literature to biology, graduating in 1929 at a time when there were few women in science. She completed an MA in marine zoology at Johns Hopkins University, taught at the University of Maryland, and published articles on natural history in the Baltimore Sun. From 1937 to 1952 she was an aquatic biologist for the U.S. Fish and Wildlife Service, resigning from her position of editor-in-chief to devote herself to writing.

Carson's unique combination of lyrical prose and accurate science earned her international literary acclaim with the publication of *The Sea Around Us* in 1951. She was awarded the National Book Award for Non-Fiction, the John Burroughs Medal, the Gold Medal of the New York Zoological Society and the Audubon Society Medal. Carson was a fellow of the Royal Society of Literature and was elected into the American Academy of Arts and Letters.

Her first book, *Under the Sea Wind*, appeared in 1941. *Silent Spring*, which alerted the world to the dangers of the misuse of pesticides, was published in 1962. Carson's articles on natural history appeared in the *Atlantic Monthly*, the *New Yorker*, *Reader's Digest* and *Holiday*. An ardent ecologist and preservationist, Carson warned against the dumping of atomic waste at sea and predicted global warming. *The Edge of the Sea*, which completed her biography of the sea, is also published in Penguin Twentieth-Century Classics. Rachel Carson died of cancer at the age of 56.

Linda Lear is Research Professor of Environmental History at George Washington University. She is the author of *Rachel Carson: Witness for Nature*, published in Penguin, and the editor of *Lost Woods: The Discovered Writing of Rachel Carson* (Beacon, 1998). She is at work on a new biography of a British writer and naturalist, and lives in Bethesda, Maryland.

Rachel Carson

Silent Spring

Introduction by Lord Shackleton
Preface by Julian Huxley, F.R.S.
With a new Afterword by Linda Lear

Penguin Books
in association with Hamish Hamilton

PENGUIN BOOKS

Published by the Penguin Group
Penguin Books Ltd, 80 Strand, London WC2R 0RL, England
Penguin Putnam Inc., 375 Hudson Street, New York, New York 10014, USA
Penguin Books Australia Ltd, 250 Camberwell Road, Camberwell, Victoria 3124, Australia
Penguin Books Canada Ltd, 10 Alcorn Avenue, Toronto, Ontario, Canada M4V 3B2
Penguin Books India (P) Ltd, 11 Community Centre, Panchsheel Park, New Delhi – 110 017, India
Penguin Books (NZ) Ltd, Cnr Rosedale and Airborne Roads, Albany, Auckland, New Zealand
Penguin Books (South Africa) (Pty) Ltd, 24 Sturdee Avenue, Rosebank 2196, South Africa

Penguin Books Ltd, Registered Offices: 80 Strand, London WC2R 0RL, England

www.penguin.com
First published in America by Houghton Mifflin 1962
First published in Great Britain by Hamish Hamilton 1963
Published in Penguin Books 1965
Reprinted in Pelican Books 1983
Reprinted in Penguin Books 1991
Reprinted with a new Afterword 1999
Reprinted in Penguin Classics 2000

038

Portions of this book were first published as a series of
articles in the *New Yorker*

Copyright © Rachel Carson, 1962
Afterword copyright © Linda Lear, 1999
All rights reserved

Set in Plantin Monotype

Printed and bound in Great Britain by Clays Ltd, Elcograf S.p.A.

ISBN-13: 978-0-141-18494-4

www.greenpenguin.co.uk

MIX
Paper from
responsible sources
FSC
www.fsc.org FSC® C018179

Penguin Books is committed to a sustainable
future for our business, our readers and our planet.
This book is made from Forest Stewardship
Council™ certified paper.

Contents

Acknowledgements 9

Author's Note 10

Introduction 11

Preface 19

1 A Fable for Tomorrow 21

2 The Obligation to Endure 23

3 Elixirs of Death 31

4 Surface Waters and Underground Seas 50

5 Realms of the Soil 61

6 Earth's Green Mantle 69

7 Needless Havoc 87

8 And No Birds Sing 100

9 Rivers of Death 122

10 Indiscriminately from the Skies 142

11 Beyond the Dreams of the Borgias 158

12 The Human Price 168

13 Through a Narrow Window 178

14 One in Every Four 193

15 Nature Fights Back 214

16 The Rumblings of an Avalanche 229

17 The Other Road 240

Afterword 258

List of Principal Sources 265

Index 309

TO ALBERT SCHWEITZER

who said

'Man has lost the capacity to foresee
and to forestall. He will end by
destroying the earth.'

The sedge is wither'd from the lake,
And no birds sing.

KEATS

I am pessimistic about the human race because it
is too ingenious for its own good. Our approach to
nature is to beat it into submission. We would
stand a better chance of survival if we accommodated ourselves to this planet and viewed it
appreciatively instead of sceptically and dictatorially.

E. B. WHITE

Acknowledgements

In a letter written in January 1958, Olga Owens Huckins told me of her own bitter experience of a small world made lifeless, and so brought my attention sharply back to a problem with which I had long been concerned. I then realized I must write this book.

During the years since then I have received help and encouragement from so many people that it is not possible to name them all here. Those who have freely shared with me the fruits of many years' experience and study represent a wide variety of government agencies in this and other countries, many universities and research institutions, and many professions. To all of them I express my deepest thanks for time and thought so generously given.

In addition my special gratitude goes to those who took time to read portions of the manuscript and to offer comment and criticism based on their own expert knowledge. Although the final responsibility for the accuracy and validity of the text is mine, I could not have completed the book without the generous help of these specialists: L. G. Bartholomew, M.D., of the Mayo Clinic, John J. Biesele of the University of Texas, A. W. A. Brown of the University of Western Ontario, Morton S. Biskind, M.D., of Westport, Connecticut, C. J. Briejèr of the Plant Protection Service in Holland, Clarence Cottam of the Rob and Bessie Welder Wildlife Foundation, George Crile, Jr, M.D., of the Cleveland Clinic, Frank Egler of Norfolk, Connecticut, Malcolm M. Hargraves, M.D., of the Mayo Clinic, W. C. Hueper, M.D., of the National Cancer Institute, C. J. Kerswill of the Fisheries Research Board of Canada, Olaus Murie of the Wilderness Society, A. D. Pickett of the Canada Department of Agriculture, Thomas G. Scott of the Illinois Natural History Survey,

Clarence Tarzwell of the Taft Sanitary Engineering Center, and George J. Wallace of Michigan State University.

Every writer of a book based on many diverse facts owes much to the skill and helpfulness of librarians. I owe such a debt to many, but especially to Ida K. Johnston of the Department of the Interior Library and to Thelma Robinson of the Library of the National Institutes of Health.

As my editor, Paul Brooks has given steadfast encouragement over the years and has cheerfully accommodated his plans to postponements and delays. For this, and for his skilled editorial judgement, I am everlastingly grateful.

I have had capable and devoted assistance in the enormous task of library research from Dorothy Algire, Jeanne Davis, and Bette Haney Duff. And I could not possibly have completed the task, under circumstances sometimes difficult, except for the faithful help of my housekeeper, Ida Sprow.

Finally, I must acknowledge our vast indebtedness to a host of people, many of them unknown to me personally, who have nevertheless made the writing of this book seem worth while. These are the people who first spoke out against the reckless and irresponsible poisoning of the world that man shares with all other creatures, and who are even now fighting the thousands of small battles that in the end will bring victory for sanity and common sense in our accommodation to the world that surrounds us.

RACHEL CARSON

AUTHOR'S NOTE

I have not wished to burden the text with footnotes but I realize that many of my readers will wish to pursue some of the subjects discussed. I have therefore included a list of my principal sources of information, arranged by chapter and page, in an appendix which will be found at the back of the book.

R.C.

Introduction

In this brilliant and controversial book, Miss Rachel Carson brings her training as a biologist and her skill as a writer to bear with great force on a significant and even sinister aspect of man's technological progress. This is the story of the use of toxic chemicals in the countryside and of the widespread destruction of wildlife in America (caused by pesticides, fungicides and herbicides). But *Silent Spring* is not merely about poisons; it is about ecology or the relation of plants and animals to their environment and to one another. Ecologists are more and more coming to recognize that for this purpose man is an animal and indeed the most important of all animals and that however artificial his dwelling, he cannot with impunity allow the natural environment of living things from which he has so recently emerged to be destroyed. Fundamentally, therefore, Miss Carson makes a well-reasoned and persuasive case for human beings to learn to appreciate the fact that they are part of the entire living world inhabiting this planet, and that they must understand its conditions of existence and so behave that these conditions are not violated.

We in Britain have not yet been exposed to the same intensity of attack as in America, but here too there is a grim side to the story. There have been, for example, the reports of a mysterious illness affecting foxes. The first substantial records of the 'fox death' were in November 1959 from near Oundle, in Northamptonshire, and soon reports were coming in from all over the country until it was estimated that 1,300 foxes had been found dead. There was much speculation as to the cause. It was suggested that death was due to a virus disease. The symptoms were striking. Foxes appeared dazed, partially blind, hypersensitive to noise, almost dying of thirst, and then death came. One odd symptom, as the Nature Conservancy reported, was that sick foxes appeared to lose their fear of mankind and were even to be found in such unlikely localities as the yard belonging to the Master of the Heythrop Hunt. No simple tests could at the time reveal the answer, but on the basis of more searching methods recently developed, 'fox death' is now generally believed to have been caused by the chlorinated hydrocarbons and other poisons so freely used in the countryside.

It was, however, the heaps of dead birds which revealed the truth. For many years biologists had given warning of danger, and already in 1960 voices were raised in Parliament and else-where demanding restriction and even a ban on chemicals such as dieldrin, aldrin, and heptachlor. It was clear that control over their use was quite inadequate and appeals were made by official bodies for more care. Then came the spring of 1961, when tens of thousands of birds were found littering the countryside, dead or dying in agony. The story from one estate alone reveals the nature of the tragedy. In the spring of 1960 at Tumby in Lincoln-shire heavy losses of birds were reported. In 1961 over 6,000 dead birds were counted. From the royal estate at Sandringham in Norfolk the list of dead birds included pheasants, red-legged partridges, partridges, woodpigeons and stock doves, green-finches, chaffinches, blackbirds, song thrushes, skylarks, moor-hens, bramblings, tree sparrows, house sparrows, jays, yellow-hammers, hedge sparrows, carrion crows, hooded crows, gold-finches, and sparrowhawks. Over 142 bodies were collected in 11½ hours of special survey counts, and hundreds more over a period of weeks. Amongst these birds were some, such as the bramblings, which are specially protected by law, yet all went down before the indiscriminate scythe of toxic chemicals.

Following this catastrophe, further pressure was brought to bear. The matter was urgently debated in Parliament. The Ministry of Agriculture, Fisheries and Food called meetings, the Nature Conservancy, backed by naturalist societies such as the Royal Society for the Protection of Birds, the British Trust for Ornithology and the Game Research Association, intervened and finally a voluntary agreement was made to refrain from using certain seed dressings, except when an attack of wheat bulb fly was seriously anticipated, and then only for autumn sowings. But there is evidence that the poisoning from sprays still goes on, though undoubtedly the voluntary ban has led to a marked reduction in the number of bird deaths caused by toxic seed dressings. Sowing conditions were particularly favourable in 1961-2 which must have had an effect in reducing the casualty figures, yet many deaths were reported from widely separated places. Once again the death roll was heavy at Tumby, especially of pheasants where the fertility of the surviving birds was seriously affected. Nest desertions began earlier in the year and out of a sample of 740 pheasants' eggs, the number hatched was well below the normal and many of the chicks were small and soon died. With the improved methods of analysis, it was found that in many of the unhatched eggs, there were present mercury and BHC (benzene hexachloride), both widely used as agricultural chemicals.

The story of the peregrine is particularly significant. It is typical of the change in our countryside which is being wrought by toxic chemicals. The peregrine, with other predators, has an important role to play in the ecology of the countryside. If you look at a map of the distribution of the peregrine in 1962 you will see that it has largely disappeared from the south of England. In the north of England peregrines are still present in fair numbers but although some pairs laid eggs, more than half of these failed. The position is similar in southern Scotland. Only in the highlands and islands has there been a fairly normal nesting season. Investigation of an egg taken from an abandoned nest near Perth showed that here again was poison.

Other predators, such as owls, have also been found dead. A significant example was that of a tawny owl from Kensington found dead on 9 July 1962. The bird was analysed by the Royal Society for the Protection of Birds' chemist and it was found to contain mercury, benzene hexachloride, heptachlor, and dieldrin. The tawny owl may well have been contaminated from eating rodents or insects in the gardens of London. A song thrush was also found dead in central London in the summer of 1962 with similar compounds in it. The number of garden chemicals on sale based on chlorinated hydrocarbons which are labelled 'safe' is a new and worrying factor, especially when one realizes that some of these contain chemicals similar to those that have wrought such havoc in the fields. It is possible that even our gardens are becoming extremely dangerous places for wild-life.

In this country there have been no great government agencies spraying whole counties and States as in America against the fire ant, the spruce bud worm or the gipsy moth and in the process seriously damaging not only wildlife but even killing domestic animals. The nearest we came to it was in the 1950s when commercial interests tried to persuade British highway authorities to switch over to the widespread use of herbicide sprays on roadside verges and hedgerows. The horrible consequence of this is well described from American experience by Rachel Carson, but in this country the Nature Conservancy, backed by enraged naturalists, managed to insist on a standstill, except for experimental treatments. Both scientific tests and cost analysis showed that inflated claims and unsubstantiated requirements for mass chemicals would not stand up to examination and, therefore, the British wayfarer and taxpayer has been spared the outrages recorded in *Silent Spring*, although strictly limited spraying on main roads here is now permitted.

The human side is perhaps the most sinister part of this book and here I must leave it to Miss Carson to tell her own very thorough story. The fact is that chemical residues are to be found in the food we eat. We are told officially that there is no hazard but we are also told by Professor Boyland, of the Chester Beatty Institute, that there is no safe dose for a carcinogen and, if there was, we would not know what it was. We are eating these chemicals, possibly in small, possibly in large quantities, and certainly they are being stored in our livers and our fat. Whether or not the evidence contained in Miss Carson's fully documented story is accepted, the fact remains that until a thing can be shown to be positively safe, we ought to reckon that any contaminant should be avoided. No one would suggest spraying fields with radioactivity, yet we do not pause before using mutagenic chemicals, the effects of which have in certain respects been shown by Dr Alexander, also of the Chester Beatty Institute, to be the same. This is no simple matter, for there are already many chemicals added to our food and there are some contaminants that occur in nature which can be dangerous to human beings.

It would be unfair to suggest that there is complete indifference in official quarters in Britain. Bodies like the British Industrial Biological Research Association have been recently set up and are actively concerned with this problem. There are high-powered Government and scientific committees and the Ministry of Agriculture, Fisheries and Food, however bland its public face, now exercises effective control to prevent the poisoning of agricultural workers and is doing a good deal more work in other parts of the field than it is generally given credit for. The same is true also of the chemical companies.

While we need to look at both sides of the coin, to remember such disasters as the Irish potato famine, yet there is a feeling of lack of urgency about the dangers, especially the hidden ones, in the use of certain poisons. The agricultural Establishment is so convinced of the great benefit in increased production through the use of these chemicals that when they come to balance the problem in utilitarian terms, they find it difficult to see the wider and longer-term consequences. It looks as if we will go on swallowing these chemicals whether we like it or not and their real effect may not be seen for another twenty or thirty years.

Nor is anything like enough research being done. This was clearly revealed in the report of the Sanders Committee. Are the gains to mankind such that we should continue to take a risk which admittedly many experts, but certainly not all, regard as negligible and, if so, are we prepared to ignore the destruction of wildlife and the cruelty? Here there is another danger and one that the ecologist is particularly aware of. Some years ago a serious plague attacked the cocoa crops in West Africa. It was found that the disease was caused by a virus found in a coccid protected by ants. The counter-attack was made on the ants, and the disease was reduced; but the natural balance was upset and later there was an outbreak of no less than four new insect plagues! Another chlorinated hydrocarbon, DDT, is already proving consistently less effective. There are no less than twenty-six kinds of malaria-carrying anopheles mosquito which are DDT-proof and the chemical weapons may prove to have broken in our hands.

The science of ecology teaches us that we have to understand the interaction of all living things in the environment in which we live. Fortunately in Great Britain there is an official agency, the Nature Conservancy, which exists to study the natural environment and to learn from research and experiment how to manage it and safeguard it so that there can be a harmonious coexistence between man and nature. Many people, however, look on the Conservancy as simply a body concerned with protecting birds, butterflies and wild flowers. It is urgently necessary that public opinion should understand more of the very serious and threatening problems with which such a body as the Conservancy has to deal, and *Silent Spring* will be an important means of enabling non-scientists to do so.

The soil is not an inert thing; it is full of minute living creatures and plants on which we depend. Yet we spray poison wholesale over it. The death of the predators is a warning to perhaps the greatest predator of all – mankind. Recently at the Wildlife Fund dinner in London, Prince Bernhard of the Netherlands said:

We are dreaming of conquering space. We are already preparing the conquest of the moon. But if we are going to treat other planets as we are treating our own, we had better leave the Moon, Mars and Venus strictly alone!

We are poisoning the air over our cities; we are poisoning the rivers and the seas; we are poisoning the soil itself. Some of this may be inevitable. But if we don't get together in a real and mighty effort to stop these attacks upon Mother Earth, wherever possible, we may find ourselves one day – one day soon, maybe – in a world that will be only a desert full of plastic, concrete and electronic robots. In that world there will be no more 'nature'; in that world man and a few domestic animals will be the only living creatures.

And yet, man cannot live without some measure of contact with nature. It is essential to his happiness.

I would ask those who find parts of this book not to their taste or consider that they can refute some of the arguments to see the picture as a whole. We are dealing with dangerous things and it may be too late to wait for positive evidence of danger. The tragedies of Thalidomide, of lung cancer from smoking, and many other examples, all these are a measure of the failure to foresee the risk and act quickly enough. A distinguished British ecologist said to me that he thought *Silent Spring* overstated some things now but in ten years' time or less these could be understatements.

Ideally, we should seek more profound solutions – resistant crop strains which would be a slow business to develop and, above all, ecological management to promote a natural balance which will also suit the needs of man. At present the university training in these fields is slight. This is not a soft option for the scientist nor, therefore, for mankind but it is one which we must face. It means more funds for fundamental research and perhaps less for developing new things directly for the market. The wild-life tragedy in the countryside involves ethical and aesthetic values and may bear on man's very survival. As the Duke of Edinburgh said at the Wildlife Fund dinner:

> Miners use canaries to warn them of deadly gases. It might not be a bad idea if we took the same warning from the dead birds in our countryside.

SHACKLETON

House of Lords,
London

Preface

I am very glad to have a share in introducing Rachel Carson's important book to the British public, though there is little that I can add to Lord Shackleton's excellent Introduction.

However, I would like to mention a few points. Pest-control is of course necessary and desirable, but it is an ecological matter, and cannot be handed over entirely to the chemists. The present campaign for mass chemical control, besides being fostered by the profit motive, is another symptom of our exaggeratedly technological and quantitative approach. The ecological approach, on the other hand, involves aiming at a dynamic balance, an integrated pattern of adjustment between a number of competing factors or even apparently conflicting interests.

Ecology in the service of man cannot be merely quantitative or arithmetical: it has to deal with total situations and must think in terms of quality as well as of quantity. One conflict is between the present and the future, between immediate and partial interests and the continuing interests of the entire human species. Accordingly ecology must aim not only at optimum use but also at optimum conservation of resources. Furthermore, these resources include enjoyment resources like scenery and solitude, beauty and interest, as well as material resources like food or minerals; and against the interest of food-production we have to balance other interests, like human health, watershed protection, and recreation.

Some of the most striking results of mass use of chemical pesticides in Britain are the virtual disappearance of so many butterflies (the buddleias that used to attract swarms of Red Admirals and Peacocks now harbour only an occasional Lesser Tortoiseshell or Cabbage White; and the chalk downs are almost bare of Blues). Cuckoos have become quite scarce owing to caterpillars – their staple diet – being killed. Song-birds are suffering from shortage of insect and worm food, as well as from the poisoning of what is left. Country hedgerows and road verges and meadows are losing their lovely and familiar flowers. In fact, as my brother Aldous said after reading Rachel Carson's book, we are losing half the subject-matter of English poetry.

The zeal for exterminating pests, rather than controlling them, of which Rachel Carson gives numerous examples, is another symptom of quantitative thinking. Indeed the very idea of extermination is unecological. It is almost certainly impossible to exterminate an abundant insect pest, but quite easy to exterminate non-abundant non-pests in the process.

It is not as if there were not methods of control available. Miss Carson gives a number of American examples of their success. One of the most interesting biological methods of controlling insect pests is by the release of irradiated males: these are sterile, and if present in sufficient numbers will enormously reduce the reproduction-rate.

Do not suppose that I am urging the abandonment of chemical control. We owe a great deal to the chemists who have given us methods of controlling the various pests that plague our lives. We have only to think of the value of antibiotics in controlling infectious disease, or of DDT in controlling malaria (though even here awkward and originally unforeseen consequences are cropping up in the shape of resistant strains of bacteria and mosquitoes). What I am against – and here I am sure that I speak for the great body of ecologists, naturalists, and conservationists – what I deplore is the advocacy and practice of mass chemical treatment as the main method of pest-control. On the contrary, though chemical control can be very useful, it too needs to be controlled, and should only be permitted when other methods are not available, and then under strict regulation and in relation to overall ecological planning.

In his closing paragraph Lord Shackleton refers to what is happening as a wildlife tragedy. It certainly is that; but it is also something more. It is an ecological tragedy. It is playing a big part in the process by which man is progressively ruining and destroying his own habitat. We must control the pest-controllers before the process gets out of hand.

JULIAN HUXLEY

Chapter 1

A Fable for Tomorrow

There was once a town in the heart of America where all life seemed to live in harmony with its surroundings. The town lay in the midst of a checkerboard of prosperous farms, with fields of grain and hillsides of orchards where, in spring, white clouds of bloom drifted above the green fields. In autumn, oak and maple and birch set up a blaze of colour that flamed and flickered across a backdrop of pines. Then foxes barked in the hills and deer silently crossed the fields, half hidden in the mists of the autumn mornings.

Along the roads, laurel, viburnum and alder, great ferns and wildflowers delighted the traveller's eye through much of the year. Even in winter the roadsides were places of beauty, where countless birds came to feed on the berries and on the seed heads of the dried weeds rising above the snow. The countryside was, in fact, famous for the abundance and variety of its bird life, and when the flood of migrants was pouring through in spring and autumn people travelled from great distances to observe them. Others came to fish the streams, which flowed clear and cold out of the hills and contained shady pools where trout lay. So it had been from the days many years ago when the first settlers raised their houses, sank their wells, and built their barns.

Then a strange blight crept over the area and everything began to change. Some evil spell had settled on the community: mysterious maladies swept the flocks of chickens; the cattle and sheep sickened and died. Everywhere was a shadow of death. The farmers spoke of much illness among their families. In the town the doctors had become more and more puzzled by new kinds of sickness appearing among their patients. There had been several sudden and unexplained deaths, not only among adults but even among children, who would be stricken suddenly while at play and die within a few hours.

There was a strange stillness. The birds, for example – where had they gone? Many people spoke of them, puzzled and disturbed. The feeding stations in the backyards were deserted. The few birds seen anywhere were moribund; they trembled violently and could not fly. It was a spring without voices. On the mornings that had once throbbed with the dawn chorus of robins, catbirds, doves, jays, wrens, and scores of other bird voices there was now no sound; only silence lay over the fields and woods and marsh.

On the farms the hens brooded, but no chicks hatched. The farmers complained that they were unable to raise any pigs – the litters were small and the young survived only a few days. The apple trees were coming into bloom but no bees droned among the blossoms, so there was no pollination and there would be no fruit.

The roadsides, once so attractive, were now lined with browned and withered vegetation as though swept by fire. These, too, were silent, deserted by all living things. Even the streams were now lifeless. Anglers no longer visited them, for all the fish had died.

In the gutters under the eaves and between the shingles of the roofs, a white granular powder still showed a few patches; some weeks before it had fallen like snow upon the roofs and the lawns, the fields and streams.

No witchcraft, no enemy action had silenced the rebirth of new life in this stricken world. The people had done it themselves.

This town does not actually exist, but it might easily have a thousand counterparts in America or elsewhere in the world. I know of no community that has experienced all the misfortunes I describe. Yet every one of these disasters has actually happened somewhere, and many real communities have already suffered a substantial number of them. A grim spectre has crept upon us almost unnoticed, and this imagined tragedy may easily become a stark reality we all shall know.

What has already silenced the voices of spring in countless towns in America? This book is an attempt to explain.

Chapter 2

The Obligation to Endure

The history of life on earth has been a history of interaction between living things and their surroundings. To a large extent, the physical form and the habits of the earth's vegetation and its animal life have been moulded by the environment. Considering the whole span of earthly time, the opposite effect, in which life actually modifies its surroundings, has been relatively slight. Only within the moment of time represented by the present century has one species – man – acquired significant power to alter the nature of his world.

During the past quarter-century this power has not only increased to one of disturbing magnitude but it has changed in character. The most alarming of all man's assaults upon the environment is the contamination of air, earth, rivers, and sea with dangerous and even lethal materials. This pollution is for the most part irrecoverable; the chain of evil it initiates not only in the world that must support life but in living tissues is for the most part irreversible. In this now universal contamination of the environment, chemicals are the sinister and little-recognized partners of radiation in changing the very nature of the world – the very nature of its life. Strontium 90, released through nuclear explosions into the air, comes to earth in rain or drifts down as fallout, lodges in soil, enters into the grass or corn or wheat grown there, and in time takes up its abode in the bones of a human being, there to remain until his death. Similarly, chemicals sprayed on croplands or forests or gardens lie long in soil, entering into living organisms, passing from one to another in a chain of poisoning and death. Or they pass mysteriously by underground streams until they emerge and, through the alchemy of air and sunlight, combine into new forms that kill vegetation, sicken cattle, and work unknown harm on those who drink from once-pure wells. As Albert Schweitzer has said, 'Man can hardly even recognize the devils of his own creation.'

It took hundreds of millions of years to produce the life that now inhabits the earth – aeons of time in which that developing and evolving and diversifying life reached a state of adjustment and balance with its surroundings. The environment, rigorously shaping and directing the life it supported, contained elements that were hostile as well as supporting. Certain rocks gave out dangerous radiation; even within the light of the sun, from which all life draws its energy, there were short-wave radiations with power to injure. Given time – time not in years but in millennia – life adjusts, and a balance has been reached. For time is the essential ingredient; but in the modern world there is no time.

The rapidity of change and the speed with which new situations are created follow the impetuous and heedless pace of man rather than the deliberate pace of nature. Radiation is no longer merely the background radiation of rocks, the bombardment of cosmic rays, the ultra-violet of the sun that have existed before there was any life on earth; radiation is now the unnatural creation of man's tampering with the atom. The chemicals to which life is asked to make its adjustment are no longer merely the calcium and silica and copper and all the rest of the minerals washed out of the rocks and carried in rivers to the sea; they are the synthetic creations of man's inventive mind, brewed in his laboratories, and having no counterparts in nature.

To adjust to these chemicals would require time on the scale that is nature's; it would require not merely the years of a man's life but the life of generations. And even this, were it by some miracle possible, would be futile, for the new chemicals come from our laboratories in an endless stream; almost five hundred annually find their way into actual use in the United States alone. The figure is staggering and its implications are not easily grasped – five hundred new chemicals to which the bodies of men and animals are required somehow to adapt each year, chemicals totally outside the limits of biologic experience.

Among them are many that are used in man's war against nature. Since the mid 1940s over two hundred basic chemicals have been created for use in killing insects, weeds, rodents, and other organisms described in the modern vernacular as 'pests'; and they are sold under several thousand different brand names.

These sprays, dusts and aerosols are now applied almost universally to farms, gardens forests, and homes – non-selective chemicals that have the power to kill every insect, the 'good' and the 'bad', to still the song of birds and the leaping of fish in the streams, to coat the leaves with a deadly film, and to linger on in soil – all this though the intended target may be only a few weeds or insects. Can anyone believe it is possible to lay down such a barrage of poisons on the surface of the earth without making it unfit for all life? They should not be called 'insecticides', but 'biocides'.

The whole process of spraying seems caught up in an endless spiral. Since DDT was released for civilian use, a process of escalation has been going on in which ever more toxic materials must be found. This has happened because insects, in a triumphant vindication of Darwin's principle of the survival of the fittest, have evolved super races immune to the particular insecticide used, hence a deadlier one has always to be developed – and then a deadlier one than that. It has happened also because, for reasons to be described later, destructive insects often undergo a 'flareback', or resurgence, after spraying, in numbers greater than before. Thus the chemical war is never won, and all life is caught in its violent crossfire.

Along with the possibility of the extinction of mankind by nuclear war, the central problem of our age has therefore become the contamination of man's total environment with such substances of incredible potential for harm – substances that accumulate in the tissues of plants and animals and even penetrate the germ cells to shatter or alter the very material of heredity upon which the shape of the future depends.

Some would-be architects of our future look towards a time when it will be possible to alter the human germ plasm by design. But we may easily be doing so now by inadvertence, for many chemicals, like radiation, bring about gene mutations. It is ironic to think that man might determine his own future by something so seemingly trivial as the choice of an insect spray.

All this has been risked – for what? Future historians may well be amazed by our distorted sense of proportion. How could intelligent beings seek to control a few unwanted species by a method that contaminated the entire environment and brought the threat of disease and death even to their own kind? Yet this is precisely what we have done. We have done it, moreover, for reasons that collapse the moment we examine them. We are told that the enormous and expanding use of pesticides is necessary to maintain farm production. Yet is our real problem not one of *over-production*? Our farms, despite measures to remove acreages from production and to pay farmers *not* to produce, have yielded such a staggering excess of crops that the American taxpayer in 1962 is paying out more than one billion dollars a year as the total carrying cost of the surplus-food storage programme. And the situation is not helped when one branch of the Agriculture Department tries to reduce production while another states, as it did in 1958,

It is believed generally that reduction of crop acreages under provisions of the Soil Bank will stimulate interest in use of chemicals to obtain maximum production on the land retained in crops.

All this is not to say there is no insect problem and no need of control. I am saying, rather, that control must be geared to realities, not to mythical situations, and that the methods employed must be such that they do not destroy us along with the insects.

The problem whose attempted solution has brought such a train of disaster in its wake is an accompaniment of our modern way of life. Long before the age of man, insects inhabited the earth — a group of extraordinarily varied and adaptable beings. Over the course of time since man's advent, a small percentage of the more than half a million species of insects have come into conflict with human welfare in two principal ways: as competitors for the food supply and as carriers of human disease.

Disease-carrying insects become important where human beings are crowded together, especially under conditions where sanitation is poor, as in time of natural disaster or war or in situations of extreme poverty and deprivation. Then control of some sort becomes necessary. It is a sobering fact, however, as we shall presently see, that the method of massive chemical control has had only limited success, and also threatens to worsen the very conditions it is intended to curb.

Under primitive agricultural conditions the farmer had few insect problems. These arose with the intensification of agriculture – the devotion of immense acreages to a single crop. Such a system set the stage for explosive increases in specific insect populations. Single-crop farming does not take advantage of the principles by which nature works; it is agriculture as an engineer might conceive it to be. Nature has introduced great variety into the landscape, but man has displayed a passion for simplifying it. Thus he undoes the built-in checks and balances by which nature holds the species within bounds. One important natural check is a limit on the amount of suitable habitat for each species. Obviously then, an insect that lives on wheat can build up its population to much higher levels on a farm devoted to wheat than on one in which wheat is intermingled with other crops to which the insect is not adapted.

The same thing happens in other situations. A generation or more ago, the towns of large areas of the United States lined their streets with the noble elm tree. Now the beauty they hopefully created is threatened with complete destruction as disease sweeps through the elms, carried by a beetle that would have only a limited chance to build up large populations and to spread from tree to tree if the elms were only occasional trees in a richly diversified planting.

Another factor in the modern insect problem is one that must be viewed against a background of geologic and human history: the spreading of thousands of different kinds of organisms from their native homes to invade new territories. This world-wide migration has been studied and graphically described by the British ecologist Charles Elton in his recent book *The Ecology of Invasions*. During the Cretaceous Period, some hundred million years ago, flooding seas cut many land bridges between continents and living things found themselves confined in what Elton calls 'colossal separate nature reserves'. There, isolated from others of their kind, they developed many new species. When some of the land masses were joined again, about fifteen million years ago, these species began to move out into new territories – a movement that is not only still in progress but is now receiving considerable assistance from man.

The importation of plants is the primary agent in the modern spread of species, for animals have almost invariably gone along with the plants, quarantine being a comparatively recent and not completely effective innovation. The United States Office of Plant Introduction alone has introduced almost 200,000 species and varieties of plants from all over the world. Nearly half of the 180 or so major insect enemies of plants in the United States are accidental imports from abroad, and most of them have come as hitch-hikers on plants.

In new territory, out of reach of the restraining hand of the natural enemies that kept down its numbers in its native land, an invading plant or animal is able to become enormously abundant. Thus it is no accident that our most troublesome insects are introduced species.

These invasions, both the naturally occurring and those dependent on human assistance, are likely to continue indefinitely. Quarantine and massive chemical campaigns are only extremely expensive ways of buying time. We are faced, according to Dr Elton, 'with a life-and-death need not just to find new technological means of suppressing this plant or that animal'; instead we need the basic knowledge of animal populations and their relations to their surroundings that will 'promote an even balance and damp down the explosive power of outbreaks and new invasions'.

Much of the necessary knowledge is now available but we do not use it. We train ecologists in our universities and even employ them in our governmental agencies but we seldom take their advice. We allow the chemical death rain to fall as though there were no alternative, whereas in fact there are many, and our ingenuity could soon discover many more if given opportunity.

Have we fallen into a mesmerized state that makes us accept as inevitable that which is inferior or detrimental, as though having lost the will or the vision to demand that which is good? Such thinking, in the words of the ecologist Paul Shepard,

idealizes life with only its head out of water, inches above the limits of toleration of the corruption of its own environment.... Why should we tolerate a diet of weak poisons, a home in insipid surroundings, a circle of acquaintances who are not quite our enemies, the noise of motors with just enough relief to prevent insanity? Who would want to live in a world which is just not quite fatal?

Yet such a world is pressed upon us. The crusade to create a chemically sterile, insect-free world seems to have engendered a fanatic zeal on the part of many specialists and most of the so-called control agencies. On every hand there is evidence that those engaged in spraying operations exercise a ruthless power. 'The regulatory entomologists ... function as prosecutor, judge and jury, tax assessor and collector and sheriff to enforce their own orders,' said Connecticut entomologist Neely Turner. The most flagrant abuses go unchecked in both state and federal agencies.

It is not my contention that chemical insecticides must never be used. I do contend that we have put poisonous and biologically potent chemicals indiscriminately into the hands of persons largely or wholly ignorant of their potentials for harm. We have subjected enormous numbers of people to contact with these poisons, without their consent and often without their knowledge. If the Bill of Rights contains no guarantee that a citizen shall be secure against lethal poisons distributed either by private individuals or by public officials, it is surely only because our forefathers, despite their considerable wisdom and foresight, could conceive of no such problem.

I contend, furthermore, that we have allowed these chemicals to be used with little or no advance investigation of their effect on soil, water, wildlife, and man himself. Future generations are unlikely to condone our lack of prudent concern for the integrity of the natural world that supports all life.

There is still very limited awareness of the nature of the threat. This is an era of specialists, each of whom sees his own problem and is unaware of or intolerant of the larger frame into which it fits. It is also an era dominated by industry, in which the right to make a dollar at whatever cost is seldom challenged. When the public protests, confronted with some obvious evidence of damaging results of pesticide applications, it is fed little tranquillizing pills of half truth. We urgently need an end to these false assurances, to the sugar coating of unpalatable facts. It is the public that is being asked to assume the risks that

the insect controllers calculate. The public must decide whether
it wishes to continue on the present road, and it can do so
only when in full possession of the facts. In the words of
Jean Rostand, 'The obligation to endure gives us the right to
know.'

Chapter 3

Elixirs of Death

For the first time in the history of the world, every human being is now subjected to contact with dangerous chemicals, from the moment of conception until death. In the less than two decades of their use, the synthetic pesticides have been so thoroughly distributed throughout the animate and inanimate world that they occur virtually everywhere. They have been recovered from most of the major river systems and even from streams of ground-water flowing unseen through the earth. Residues of these chemicals linger in soil to which they may have been applied a dozen years before. They have entered and lodged in the bodies of fish, birds, reptiles, and domestic and wild animals so universally that scientists carrying on animal experiments find it almost impossible to locate subjects free from such contamination. They have been found in fish in remote mountain lakes, in earthworms burrowing in soil, in the eggs of birds – and in man himself. For these chemicals are now stored in the bodies of the vast majority of human beings, regardless of age. They occur in the mother's milk, and probably in the tissues of the unborn child.

All this has come about because of the sudden rise and prodigious growth of an industry for the production of man-made or synthetic chemicals with insecticidal properties. This industry is a child of the Second World War. In the course of developing agents of chemical warfare, some of the chemicals created in the laboratory were found to be lethal to insects. The discovery did not come by chance: insects were widely used to test chemicals as agents of death for man.

The result has been a seemingly endless stream of synthetic insecticides. In being man-made – by ingenious laboratory manipulation of the molecules, substituting atoms, altering their arrangement – they differ sharply from the simpler inorganic insecticides of pre-war days. These were derived from naturally occurring minerals and plant products – compounds of arsenic, copper, lead, manganese, zinc, and other minerals, pyrethrum from the dried flowers of chrysanthemums, nicotine sulphate from some of the relatives of tobacco, and rotenone from leguminous plants of the East Indies.

What sets the new synthetic insecticides apart is their enormous biological potency. They have immense power not merely to poison but to enter into the most vital processes of the body and change them in sinister and often deadly ways. Thus, as we shall see, they destroy the very enzymes whose function is to protect the body from harm, they block the oxidation processes from which the body receives its energy, they prevent the normal functioning of various organs, and they may initiate in certain cells the slow and irreversible change that leads to malignancy.

Yet new and more deadly chemicals are added to the list each year and new uses are devised so that contact with these materials has become practically world-wide. The production of synthetic pesticides in the United States soared from 124,259,000 pounds in 1947 to 637,666,000 pounds in 1960 – more than a fivefold increase. The wholesale value of these products was well over a quarter of a billion dollars. But in the plans and hopes of the industry this enormous production is only a beginning.

A Who's Who of pesticides is therefore of concern to us all. If we are going to live so intimately with these chemicals – eating and drinking them, taking them into the very marrow of our bones – we had better know something about their nature and their power.

Although the Second World War marked a turning away from inorganic chemicals as pesticides into the wonder world of the carbon molecule, a few of the old materials persist. Chief among these is arsenic, which is still the basic ingredient in a variety of weed and insect killers. Arsenic is a highly toxic mineral occurring widely in association with the ores of various metals, and in very small amounts in volcanoes, in the sea, and in spring water. Its relations to man are varied and historic. Since many of its compounds are tasteless, it has been a favourite agent of homicide from long before the time of the Borgias to the present. Arsenic was the first recognized elementary carcinogen (or cancer-causing substance), identified in chimney soot and linked to cancer nearly two centuries ago by an English physician. Epidemics of chronic arsenical poisoning involving whole populations over long periods are on record. Arsenic-contaminated environments have also caused sickness and death among horses, cows, goats, pigs, deer, fishes, and bees; despite this record arsenical sprays and dusts are widely used. In the arsenic-sprayed cotton country of southern United States, beekeeping as an industry has nearly died out. Farmers using arsenic dusts over long periods have been afflicted with chronic arsenic poisoning; livestock have been poisoned by crop sprays or weed killers containing arsenic. Drifting arsenic dusts from blueberry lands have spread over neighbouring farms, contaminating streams, fatally poisoning bees and cows, and causing human illness.

It is scarcely possible ... to handle arsenicals with more utter disregard of the general health than that which has been practised in our country in recent years [said Dr W. C. Hueper, of the National Cancer Institute, an authority on environmental cancer]. Anyone who has watched the dusters and sprayers of arsenical insecticides at work must have been impressed by the almost supreme carelessness with which the poisonous substances are dispensed.

Modern insecticides are still more deadly. The vast majority fall into one of two large groups of chemicals. One, represented by DDT, is known as the 'chlorinated hydrocarbons'. The other group consists of the organic phosphorus insecticides, and is represented by the reasonably familiar malathion and parathion. All have one thing in common. As mentioned above, they are built on a basis of carbon atoms, which are also the indispensable building blocks of the living world, and thus classed as 'organic'. To understand them, we must see of what they are made, and how, although linked with the basic chemistry of all life, they lend themselves to the modifications which make them agents of death.

The basic element, carbon, is one whose atoms have an almost infinite capacity for uniting with each other in chains and rings and various other configurations, and for becoming linked with atoms of other substances. Indeed, the incredible diversity of living creatures from bacteria to the great blue whale is largely due to this capacity of carbon. The complex protein molecule has the carbon atom as its basis, as have molecules of fat, carbo-hydrates, enzymes, and vitamins. So, too, have enormous num-bers of non-living things, for carbon is not necessarily a symbol of life.

Some organic compounds are simply combinations of carbon and hydrogen. The simplest of these is methane, or marsh gas, formed in nature by the bacterial decomposition of organic matter under water. Mixed with air in proper proportions, methane becomes the dreaded 'fire damp' of coal mines. Its structure is beautifully simple, consisting of one carbon atom to which four hydrogen atoms have become attached:

Chemists have discovered that it is possible to detach one or all of the hydrogen atoms and substitute other elements. For ex-ample, by substituting one atom of chlorine for one of hydrogen we produce methyl chloride:

Take away three hydrogen atoms and substitute chlorine and we have the anaesthetic chloroform:

Substitute chlorine atoms for all of the hydrogen atoms and the result is carbon tetrachloride, the familiar cleaning fluid:

In the simplest possible terms, these changes rung upon the basic molecule of methane illustrate what a chlorinated hydrocarbon is. But this illustration gives little hint of the true complexity of the chemical world of the hydrocarbons, or of the manipulations by which the organic chemist creates his infinitely varied materials. For instead of the simple methane molecule with its single carbon atom, he may work with hydrocarbon molecules consisting of many carbon atoms, arranged in rings or chains, with side chains or branches, holding to themselves with chemical bonds not merely simple atoms of hydrogen or chlorine but also a wide variety of chemical groups. By seemingly slight changes the whole character of the substance is changed; for example, not only what is attached but the place of attachment to the carbon atom is highly important. Such ingenious manipulations have produced a battery of poisons of truly extraordinary power.

DDT (short for dichloro-diphenyl-trichloro-ethane) was first synthesized by a German chemist in 1874, but its properties as an insecticide were not discovered until 1939. Almost immediately DDT was hailed as a means of stamping out insect-borne disease and winning the farmers' war against crop destroyers overnight. The discoverer, Paul Müller of Switzerland, won the Nobel Prize.

DDT is now so universally used that in most minds the product takes on the harmless aspect of the familiar. Perhaps the myth of the harmlessness of DDT rests on the fact that one of its first uses was the wartime dusting of many thousands of soldiers, refugees, and prisoners, to combat lice. It is widely believed that since so many people came into extremely intimate contact with DDT and suffered no immediate ill effects the chemical must certainly be innocent of harm. This understandable misconception arises from the fact that – unlike other chlorinated hydrocarbons – DDT *in powder form* is not readily absorbed through the skin. Dissolved in oil, as it usually is, DDT is definitely toxic. If swallowed, it is absorbed slowly through the digestive tract; it may also be absorbed through the lungs. Once it has entered the body it is stored largely in organs rich in fatty substances (because DDT itself is fat-soluble) such as the adrenals, testes, or thyroid. Relatively large amounts are deposited in the liver, kidneys, and the fat of the large, protective mesenteries that enfold the intestines.

This storage of DDT begins with the smallest conceivable intake of the chemical (which is present as residues on most foodstuffs) and continues until quite high levels are reached. The fatty storage depots act as biological magnifiers, so that an intake of as little as $\frac{1}{10}$ of 1 part per million in the diet results in storage of about 10 to 15 parts per million, an increase of one hundredfold or more. These terms of reference, so commonplace to the chemist or the pharmacologist, are unfamiliar to most of us. One part in a million sounds like a very small amount – and so it is. But such substances are so potent that a minute quantity can bring about vast changes in the body. In animal experiments, 3 parts per million has been found to inhibit an essential enzyme in heart muscle; only 5 parts per million has brought about necrosis or disintegration of liver cells; only 2·5 parts per million of the closely related chemicals dieldrin and chlordane did the same.

This is really not surprising. In the normal chemistry of the human body there is just such a disparity between cause and effect. For example, a quantity of iodine as small as two tenthousandths of a gram spells the difference between health and disease. Because these small amounts of pesticides are cumulatively stored and only slowly excreted, the threat of chronic poisoning and degenerative changes of the liver and other organs is very real.

Scientists do not agree upon how much DDT can be stored in the human body. Dr Arnold Lehman, who is the chief pharmacologist of the Food and Drug Administration, says there is neither a floor below which DDT is not absorbed nor a ceiling beyond which absorption and storage ceases. On the other hand, Dr Wayland Hayes of the United States Public Health Service contends that in every individual a point of equilibrium is reached, and that DDT in excess of this amount is excreted. For practical purposes it is not particularly important which of these men is right. Storage in human beings has been well investigated, and we know that the average person is storing potentially harmful amounts. According to various studies, individuals with no known exposure (except the inevitable dietary one) store an average of 5·3 parts per million to 7·4 parts per million; agricultural workers 17·1 parts per million; and workers in insecticide plants as high as 648 parts per million! So the range of proven storage is quite wide and, what is even more to the point, the minimum figures are above the level at which damage to the liver and other organs or tissues may begin.

One of the most sinister features of DDT and related chemicals is the way they are passed on from one organism to another through all the links of the food chains. For example, fields of alfalfa are dusted with DDT; meal is later prepared from the alfalfa and fed to hens; the hens lay eggs which contain DDT. Or the hay, containing residues of 7 to 8 parts per million, may be fed to cows. The DDT will turn up in the milk in the amount of about 3 parts per million, but in butter made from this milk the concentration may run to 65 parts per million. Through such a process of transfer, what started out as a very small amount of DDT may end as a heavy concentration. Farmers nowadays find it difficult to obtain uncontaminated fodder for their milk cows, though the Food and Drug Administration forbids the presence of insecticide residues in milk shipped in inter-state commerce.

The poison may also be passed on from mother to offspring. Insecticide residues have been recovered from human milk in samples tested by Food and Drug Administration scientists. This means that the breast-fed human infant is receiving small but regular additions to the load of toxic chemicals building up in his body. It is by no means his first exposure, however: there is good reason to believe this begins while he is still in the womb. In experimental animals the chlorinated hydrocarbon insecticides freely cross the barrier of the placenta, the traditional protective shield between the embryo and harmful substances in the mother's body. While the quantities so received by human infants would normally be small, they are not unimportant because children are more susceptible to poisoning than adults. This situation also means that today the average individual almost certainly starts life with the first deposit of the growing load of chemicals his body will be required to carry thenceforth.

All these facts – storage at even low levels, subsequent accumulation, and occurrence of liver damage at levels that may easily occur in normal diets – caused Food and Drug Administration scientists to declare as early as 1950 that it is 'extremely likely the potential hazard of DDT has been under-estimated'. There has been no such parallel situation in medical history. No one yet knows what the ultimate consequences may be.

Chlordane, another chlorinated hydrocarbon, has all the unpleasant attributes of DDT plus a few that are peculiarly its own. Its residues are long persistent in soil, on foodstuffs, or on surfaces to which it may be applied, yet it is also quite volatile and poisoning by inhalation is a definite risk to anyone handling or exposed to it. Chlordane makes use of all available portals to enter the body. It penetrates the skin easily, is breathed in as vapour, and of course is absorbed from the digestive tract if residues are swallowed. Like all other chlorinated hydrocarbons, its deposits build up in the body in cumulative fashion. A diet containing such a small amount of chlordane as 2·5 parts per million may eventually lead to storage of 75 parts per million in the fat of experimental animals.

So experienced a pharmacologist as Dr Lehman has described chlordane as 'one of the most toxic of insecticides – anyone handling it could be poisoned'. Judging by the carefree liberality with which dusts for lawn treatments by suburbanites are laced with chlordane, this warning has not been taken to heart. The fact that the suburbanite is not instantly stricken has little meaning, for the toxins may sleep long in his body, to become manifest months or years later in an obscure disorder almost impossible to trace to its origins. On the other hand, death may strike quickly. One victim who accidentally spilled a 25 per cent solution on his skin developed symptoms of poisoning within forty minutes and died before medical help could be obtained.

No reliance can be placed on receiving advance warning which might allow treatment to be had in time.

Heptachlor, one of the constituents of chlordane, is marketed as a separate formulation. It has a particularly high capacity for storage in fat. If the diet contains as little as $\frac{1}{10}$ of 1 part per million there will be measurable amounts of heptachlor in the body. It also has the curious ability to undergo change into a chemically distinct substance known as heptachlor epoxide. It does this in soil and in the tissues of both plants and animals. Tests on birds indicate that the epoxide that results from this change is about four times as toxic as the original chemical, which in turn is four times as toxic as chlordane.

As long ago as the mid 1930s a special group of hydrocarbons, the chlorinated naphthalenes, was found to cause hepatitis, and also a rare and almost invariably fatal liver disease in persons subjected to occupational exposure. They have led to illness and death of workers in electrical industries; and more recently, in agriculture, they have been considered a cause of a mysterious and usually fatal disease of cattle. In view of these antecedents, it is not surprising that three of the insecticides that belong to this group are among the most violently poisonous of all the hydrocarbons. These are dieldrin, aldrin, and endrin.

Dieldrin, named after a German chemist, Diels, is about five times as toxic as DDT when swallowed but forty times as toxic when absorbed through the skin in solution. It is notorious for striking quickly and with terrible effect at the nervous system, sending the victims into convulsions. Persons thus poisoned recover so slowly as to indicate chronic effects. As with other chlorinated hydrocarbons, these long-term effects include severe damage to the liver. The long duration of its residues and the effective insecticidal action make dieldrin one of the most used insecticides today, despite the appalling destruction of wildlife that has followed its use. As tested on quail and pheasants, it has proved to be about forty or fifty times as toxic as DDT.

There are vast gaps in our knowledge of how dieldrin is stored or distributed in the body, or excreted, for the chemists' ingenuity in devising insecticides has long ago outrun biological knowledge of the way these poisons affect the living organism. However, there is every indication of long storage in the human body, where deposits may lie dormant like a slumbering volcano, only to flare up in periods of physiological stress when the body draws upon its fat reserves. Much of what we do know has been learned through hard experience in the anti-malarial campaigns carried out by the World Health Organization. As soon as dieldrin was substituted for DDT in malaria-control work (because the malaria mosquitoes had become resistant to DDT), cases of poisoning among the spraymen began to occur. The seizures were severe – from half to all (varying in the different programmes) of the men affected went into convulsions and several died. Some had convulsions as long as *four months* after the last exposure.

Aldrin is a somewhat mysterious substance, for although it exists as a separate entity it bears the relation of alter ego to dieldrin. When carrots are taken from a bed treated with aldrin they are found to contain residues of dieldrin. This change occurs in living tissues and also in soil. Such alchemistic transformations have led to many erroneous reports, for if a chemist, knowing aldrin has been applied, tests for it he will be deceived into thinking all residues have been dissipated. The residues are there, but they are dieldrin and this requires a different test.

Like dieldrin, aldrin is extremely toxic. It produces degenerative changes in the liver and kidneys. A quantity the size of an aspirin tablet is enough to kill more than four hundred quail. Many cases of human poisonings are on record, most of them in connection with industrial handling.

Aldrin, like most of this group of insecticides, projects a menacing shadow into the future, the shadow of sterility. Pheasants fed quantities too small to kill them nevertheless laid few eggs, and the chicks that hatched soon died. The effect is not confined to birds. Rats exposed to aldrin had fewer pregnancies and their young were sickly and short-lived. Puppies born of treated mothers died within three days. By one means or another, the new generations suffer for the poisoning of their parents. No one knows whether the same effect will be seen in human beings, yet this chemical has been sprayed from aeroplanes over suburban areas and farmlands.

Endrin is the most toxic of all the chlorinated hydrocarbons. Although chemically rather closely related to dieldrin, a little twist in its molecular structure makes it five times as poisonous. It makes the progenitor of all this group of insecticides, DDT, seem by comparison almost harmless. It is fifteen times as poisonous as DDT to mammals, thirty times as poisonous to fish, and about 300 times as poisonous to some birds.

In the decade of its use, endrin has killed enormous numbers of fish, has fatally poisoned cattle that have wandered into sprayed orchards, has poisoned wells, and has drawn a sharp warning from at least one state health department that its careless use is endangering human lives.

In one of the most tragic cases of endrin poisoning there was no apparent carelessness; efforts had been made to take precautions apparently considered adequate. A year-old child had been taken by his American parents to live in Venezuela. There were cockroaches in the house to which they moved, and after a few days a spray containing endrin was used. The baby and the small family dog were taken out of the house before the spraying was done about nine o'clock one morning. After the spraying the floors were washed. The baby and dog were returned to the house in mid-afternoon. An hour or so later the dog vomited, went into convulsions, and died. At 10 p.m. on the evening of the same day the baby also vomited, went into convulsions, and lost consciousness. After that fateful contact with endrin, this normal, healthy child became little more than a vegetable – unable to see or hear, subject to frequent muscular spasms, apparently completely cut off from contact with his surroundings. Several months of treatment in a New York hospital failed to change his condition or bring hope of change. 'It is extremely doubtful,' reported the attending physicians, 'that any useful degree of recovery will occur.'

The second major group of insecticides, the alkyl or organic phosphates, are among the most poisonous chemicals in the world. The chief and most obvious hazard attending their use is that of acute poisoning of people applying the sprays or accidentally coming in contact with drifting spray, with vegetation coated by it, or with a discarded container. In Florida, two children found an empty bag and used it to repair a swing. Shortly thereafter both of them died and three of their playmates became ill. The bag had once contained an insecticide called parathion, one of the organic phosphates; tests established death by parathion poisoning. On another occasion two small boys in Wisconsin, cousins, died on the same night. One had been playing in his yard when spray drifted in from an adjoining field where his father was spraying potatoes with parathion; the other had run playfully into the barn after his father and had put his hand on the nozzle of the spray equipment.

The origin of these insecticides has a certain ironic significance. Although some of the chemicals themselves – organic esters of phosphoric acid – had been known for many years, their insecticidal properties remained to be discovered by a German chemist, Gerhard Schrader, in the late 1930s. Almost immediately the German government recognized the value of these same chemicals as new and devastating weapons in man's war against his own kind, and the work on them was declared secret. Some became the deadly nerve gases. Others, of closely allied structure, became insecticides.

The organic phosphorus insecticides act on the living organism in a peculiar way. They have the ability to destroy enzymes – enzymes that perform necessary functions in the body. Their target is the nervous system, whether the victim is an insect or a warm-blooded animal. Under normal conditions, an impulse passes from nerve to nerve with the aid of a 'chemical transmitter' called acetylcholine, a substance that performs an essential function and then disappears. Indeed, its existence is so ephemeral that medical researchers are unable, without special procedures, to sample it before the body has destroyed it. This transient nature of the transmitting chemical is necessary to the normal functioning of the body. If the acetylcholine is not destroyed as soon as a nerve impulse has passed, impulses continue to flash across the bridge from nerve to nerve, as the chemical exerts its effects in an ever more intensified manner. The movements of the whole body become uncoordinated: tremors, muscular spasms, convulsions, and death quickly result.

This contingency has been provided for by the body. A protective enzyme called cholinesterase is at hand to destroy the transmitting chemical once it is no longer needed. By this means a precise balance is struck and the body never builds up a dangerous amount of acetylcholine. But on contact with the organic phosphorus insecticides, the protective enzyme is destroyed, and as the quantity of the enzyme is reduced that of the transmitting chemical builds up. In this effect, the organic phosphorus compounds resemble the alkaloid poison muscarine, found in a poisonous mushroom, the fly amanita.

Repeated exposures may lower the cholinesterase level until an individual reaches the brink of acute poisoning, a brink over which he may be pushed by a very small additional exposure. For this reason it is considered important to make periodic examinations of the blood of spray operators and others regularly exposed.

Parathion is one of the most widely used of the organic phosphates. It is also one of the most powerful and dangerous. Honey bees become 'wildly agitated and bellicose' on contact with it, perform frantic cleaning movements, and are near death within half an hour. A chemist, thinking to learn by the most direct possible means the dose acutely toxic to human beings, swallowed a minute amount, equivalent to about ·00424 oz. Paralysis followed so instantaneously that he could not reach the antidotes he had prepared at hand, and so he died. Parathion is now said to be a favourite instrument of suicide in Finland. In recent years the State of California has reported an average of more than two hundred cases of accidental parathion poisoning annually. In many parts of the world the fatality rate from parathion is startling: 100 fatal cases in India and 67 in Syria in 1958, and an average of 336 deaths per year in Japan.

Yet some 7,000,000 pounds of parathion are now applied to fields and orchards of the United States – by hand-sprayers, motorized blowers and dusters, and by aeroplane. The amount used on California farms alone could, according to one medical authority, 'provide a lethal dose for five to ten times the whole world's population'.

One of the few circumstances that save us from extinction by this means is the fact that parathion and other chemicals of this group are decomposed rather rapidly. Their residues on the crops to which they are applied are therefore relatively short-lived compared with the chlorinated hydrocarbons. However, they last long enough to create hazards and produce consequences that range from the merely serious to the fatal. In Riverside, California, eleven out of thirty men picking oranges became violently ill and all but one had to be hospitalized. Their symptoms were typical of parathion poisoning. The grove had been sprayed with parathion some two and a half weeks earlier; the residues that reduced them to retching, half-blind, semi-conscious misery were sixteen to nineteen days old. And this is not by any means a record for persistence. Similar mishaps have occurred in groves sprayed a month earlier, and residues have been found in the peel of oranges six months after treatment with standard dosages.

The danger to all workers applying the organic phosphorus insecticides in fields, orchards, and vineyards is so extreme that some states using these chemicals have established laboratories where physicians may obtain aid in diagnosis and treatment. Even the physicians themselves may be in some danger, unless they wear rubber gloves in handling the victims of poisoning. So may a laundress washing the clothing of such victims, which may have absorbed enough parathion to affect her.

Malathion, another of the organic phosphates, is almost as familiar to the public as DDT, being widely used by gardeners, in household insecticides, in mosquito spraying, and in such blanket attacks on insects as the spraying of nearly a million acres of Florida communities for the Mediterranean fruit fly. It is considered the least toxic of this group of chemicals and many people assume they may use it freely and without fear of harm. Commercial advertising encourages this comfortable attitude.

The alleged 'safety' of malathion rests on rather precarious ground, although – as often happens – this was not discovered until the chemical had been in use for several years. Malathion is 'safe' only because the mammalian liver, an organ with extra-ordinary protective powers, renders it relatively harmless. The detoxification is accomplished by one of the enzymes of the liver. If, however, something destroys this enzyme or interferes with its action, the person exposed to malathion receives the full force of the poison.

Unfortunately for all of us, opportunities for this sort of thing to happen are legion. A few years ago a team of Food and Drug Administration scientists discovered that when malathion and certain other organic phosphates are administered simultaneously a massive poisoning results – up to fifty times as severe as would be predicted on the basis of adding together the toxicities of the two. In other words, one-hundredth of the lethal dose of each compound may be fatal when the two are combined.

This discovery led to the testing of other combinations. It is now known that many pairs of organic phosphate insecticides are highly dangerous, the toxicity being stepped up or 'potentiated' through the combined action. Potentiation seems to take place when one compound destroys the liver enzyme responsible for detoxifying the other. The two need not be given simultaneously. The hazard exists not only for the man who may spray this week with one insecticide and next week with another; it exists also for the consumer of sprayed products. The common salad bowl may easily present a combination of organic phosphate insecticides. Residues well within the legally permissible limits may interact.

The full scope of the dangerous interaction of chemicals is as yet little known, but disturbing findings now come regularly from scientific laboratories. Among these is the discovery that the toxicity of an organic phosphate can be increased by a second agent that is not necessarily an insecticide. For example, one of the plasticizing agents may act even more strongly than another insecticide to make malathion more dangerous. Again, this is because it inhibits the liver enzyme that normally would 'draw the teeth' of the poisonous insecticide.

What of other chemicals in the normal human environment? What, in particular, of drugs? A bare beginning has been made on this subject, but already it is known that some organic phosphates (parathion and malathion) increase the toxicity of some drugs used as muscle relaxants, and that several others (again including malathion) markedly increase the sleeping time of barbiturates.

In Greek mythology the sorceress Medea, enraged at being supplanted by a rival for the affections of her husband Jason, presented the new bride with a robe possessing magic properties. The wearer of the robe immediately suffered a violent death. This death-by-indirection now finds its counterpart in what are known as 'systemic insecticides'. These are chemicals with extraordinary properties which are used to convert plants or animals into a sort of Medea's robe by making them actually poisonous. This is done with the purpose of killing insects that may come in contact with them, especially by sucking their juices or blood.

The world of systemic insecticides is a weird world, surpassing the imaginings of the brothers Grimm – perhaps most closely akin to the cartoon world of Charles Addams. It is a world where the enchanted forest of the fairy-tales has become the poisonous forest in which an insect that chews a leaf or sucks the sap of a plant is doomed. It is a world where a flea bites a dog, and dies because the dog's blood has been made poisonous, where an insect may die from vapours emanating from a plant it has never touched, where a bee may carry poisonous nectar back to its hive and presently produce poisonous honey.

The entomologist's dream of the built-in insecticide was born when workers in the field of applied entomology realized they could take a hint from nature: they found that wheat growing in soil containing sodium selenate was immune to attack by aphids or spider mites. Selenium, a naturally occurring element found sparingly in rocks and soils of many parts of the world, thus became the first systemic insecticide.

What makes an insecticide a systemic is the ability to permeate all the tissues of a plant or animal and make them toxic. This quality is possessed by some chemicals of the chlorinated hydrocarbon group and by others of the organophosphorus group, all synthetically produced, as well as by certain naturally occurring substances. In practice, however, most systemics are drawn from the organophosphorus group because the problem of residues is somewhat less acute.

Systemics act in other devious ways. Applied to seeds, either by soaking or in a coating combined with carbon, they extend their effects into the following plant generation and produce seedlings poisonous to aphids and other sucking insects. Vegetables such as peas, beans, and sugar beets are sometimes thus protected. Cotton seeds coated with a systemic insecticide have been in use for some time in California, where twenty-five farm labourers planting cotton in the San Joaquin Valley in 1959 were seized with sudden illness, caused by handling the bags of treated seeds.

In England someone wondered what happened when bees made use of nectar from plants treated with systemics. This was investigated in areas treated with a chemical called schradan. Although the plants had been sprayed before the flowers were formed, the nectar later produced contained the poison. The result, as might have been predicted, was that the honey made by the bees also was contaminated with schradan.

Use of animal systemics has concentrated chiefly on control of the cattle grub, a damaging parasite of livestock. Extreme care must be used in order to create an insecticidal effect in the blood and tissues of the host without setting up a fatal poisoning. The balance is delicate and government veterinarians have found that repeated small doses can gradually deplete an animal's supply of the protective enzyme cholinesterase, so that without warning a minute additional dose will cause poisoning.

There are strong indications that fields closer to our daily lives are being opened up. You may now give your dog a pill which, it is claimed, will rid him of fleas by making his blood poisonous to them. The hazards discovered in treating cattle would presumably apply to the dog. As yet no one seems to have proposed a human systemic that would make us lethal to a mosquito. Perhaps this is the next step.

So far in this chapter we have been discussing the deadly chemicals that are being used in our war against the insects. What of our simultaneous war against the weeds?

The desire for a quick and easy method of killing unwanted plants has given rise to a large and growing array of chemicals that are known as herbicides, or, less formally, as weed killers. The story of how these chemicals are used and misused will be told in Chapter 6; the question that here concerns us is whether the weed killers are poisons and whether their use is contributing to the poisoning of the environment.

The legend that the herbicides are toxic only to plants and so pose no threat to animal life has been widely disseminated, but unfortunately it is not true. The plant killers include a large variety of chemicals that act on animal tissue as well as on vegetation. They vary greatly in their action on the organism. Some are general poisons, some are powerful stimulants of metabolism, causing a fatal rise in body temperature, some induce malignant tumours either alone or in partnership with other chemicals, some strike at the genetic material of the race by causing gene mutations. The herbicides, then, like the insecticides, include some very dangerous chemicals, and their careless use in the belief they are 'safe' can have disastrous results.

Despite the competition of a constant stream of new chemicals issuing from the laboratories, arsenic compounds are still liberally used, both as insecticides (as mentioned above) and as weed killers, where they usually take the chemical form of sodium arsenite. The history of their use is not reassuring. As roadside sprays, they have cost many a farmer his cow and killed uncounted numbers of wild creatures. As aquatic weed killers in lakes and reservoirs they have made public waters unsuitable for drinking or even for swimming. As a spray applied to potato fields to destroy the vines they have taken a toll of human and non-human life.

In England this latter practice developed about 1951 as a result of a shortage of sulphuric acid, formerly used to burn off the potato vines. The Ministry of Agriculture considered it necessary to give warning of the hazard of going into the arsenic-sprayed fields, but the warning was not understood by the cattle (nor, we must assume, by the wild animals and birds) and reports of cattle poisoned by the arsenic sprays came with monotonous regularity. When death came also to a farmer's wife through arsenic-contaminated water, one of the major English chemical companies (in 1959) stopped production of arsenical sprays and called in supplies already in the hands of dealers, and shortly thereafter the Ministry of Agriculture announced that because of high risks to people and cattle restrictions on the use of arsenics would be imposed. In 1961, the Australian government announced a similar ban. No such restrictions impede the use of these poisons in the United States, however.

Some of the 'dinitro' compounds are also used as herbicides. They are rated as among the most dangerous materials of this type in use in the United States. Dinitrophenol is a strong metabolic stimulant. For this reason it was at one time used as a reducing drug, but the margin between the slimming dose and that required to poison or kill was slight – so slight that several patients died and many suffered permanent injury before use of the drug was finally halted.

A related chemical, pentachlorophenol, sometimes known as 'penta', is used as a weed killer as well as an insecticide, often being sprayed along railway tracks and in waste areas. Penta is extremely toxic to a wide variety of organisms from bacteria to man. Like the dinitros, it interferes, often fatally, with the body's source of energy, so that the affected organism almost literally burns itself up. Its fearful power is illustrated in a fatal accident recently reported by the California Department of Health. A tank truck driver was preparing a cotton defoliant by mixing diesel oil with pentachlorophenol. As he was drawing the concentrated chemical out of a drum, the spigot accidentally toppled back. He reached in with his bare hand to regain the spigot. Although he washed immediately, he became acutely ill and died the next day.

While the results of weed killers such as sodium arsenite or the phenols are grossly obvious, some other herbicides are more insidious in their effects. For example, the now famous cranberry-weed-killer aminotriazole, or amitrol, is rated as having relatively low toxicity. But in the long run its tendency to cause malignant tumours of the thyroid may be far more significant for wildlife and perhaps also for man.

Among the herbicides are some which are classified as 'muta-gens', or agents capable of modifying the genes, the materials of heredity. We are rightly appalled by the genetic effects of radiation; how, then, can we be indifferent to the same effect in chemicals that we disseminate widely in our environment?

Chapter 4

Surface Waters and
Underground Seas

Of all our natural resources water has become the most precious.
By far the greater part of the earth's surface is covered by its
enveloping seas, yet in the midst of this plenty we are in want.
By a strange paradox, most of the earth's abundant water is not
usable for agriculture, industry, or human consumption because
of its heavy load of sea salts, and so most of the world's population is either experiencing or is threatened with critical shortages.
In an age when man has forgotten his origins and is blind even
to his most essential needs for survival, water along with other
resources has become the victim of his indifference.

The problem of water pollution by pesticides can be understood only in context, as part of the whole to which it belongs –
the pollution of the total environment of mankind. The pollution
entering our waterways comes from many sources: radioactive
wastes from reactors, laboratories, and hospitals; fallout from
nuclear explosions; domestic wastes from cities and towns;
chemical wastes from factories. To these is added a new kind of
fallout – the chemical sprays applied to croplands and gardens,
forests and fields. Many of the chemical agents in this alarming
mélange imitate and augment the harmful effects of radiation,
and within the groups of chemicals themselves there are sinister
and little-understood interactions, transformations, and summa-
tions of effect.

Ever since chemists began to manufacture substances that
nature never invented, the problems of water purification have
become complex and the danger to users of water has increased.
As we have seen, the production of these synthetic chemicals in
large volume began in the 1940s. It has now reached such proportions that an appalling deluge of chemical pollution is daily
poured into the nation's waterways. When inextricably mixed
with domestic and other wastes discharged into the same water,

these chemicals sometimes defy detection by the methods in ordinary use by purification plants. Most of them are so stable that they cannot be broken down by ordinary processes. Often they cannot even be identified. In rivers, a really incredible variety of pollutants combine to produce deposits that the sanitary engineers can only despairingly refer to as 'gunk'. Professor Rolf Eliassen of the Massachusetts Institute of Technology testified before a congressional committee to the impossibility of predicting the composite effect of these chemicals, or of identifying the organic matter resulting from the mixture. 'We don't begin to know what that is,' said Professor Eliassen. 'What is the effect on the people? We don't know.'

To an ever-increasing degree, chemicals used for the control of insects, rodents, or unwanted vegetation contribute to these organic pollutants. Some are deliberately applied to bodies of water to destroy plants, insect larvae, or undesired fishes. Some come from forest spraying that may blanket two or three million acres of a single state with spray directed against a single insect pest – spray that falls directly into streams or that drips down through the leafy canopy to the forest floor, there to become part of the slow movement of seeping moisture beginning its long journey to the sea. Probably the bulk of such contaminants are the water-borne residues of the millions of pounds of agricultural chemicals that have been applied to farmlands for insect or rodent control and have been leached out of the ground by rains to become part of the universal seaward movement of water.

Here and there we have dramatic evidence of the presence of these chemicals in our streams and even in public water supplies. For example, a sample of drinking water from an orchard area in Pennsylvania, when tested on fish in a laboratory, contained enough insecticide to kill all of the test fish in only four hours. Water from a stream draining sprayed cotton fields remained lethal to fishes even after it had passed through a purifying plant, and in fifteen streams tributary to the Tennessee River in Alabama the run-off from fields treated with toxaphene, a chlorinated hydrocarbon, killed all the fish inhabiting the streams. Two of these streams were sources of municipal water supply. Yet for a week after the application of the insecticide the water remained poisonous, a fact attested by the daily deaths of gold-fish suspended in cages downstream.

For the most part this pollution is unseen and invisible, making its presence known when hundreds or thousands of fish die, but more often never detected at all. The chemist who guards water purity has no routine tests for these organic pollutants and no way to remove them. But whether detected or not, the pesticides are there, and as might be expected with any materials applied to land surfaces on so vast a scale, they have now found their way into many and perhaps all of the major river systems of the country.

If anyone doubts that our waters have become almost univer-sally contaminated with insecticides he should study a small report issued by the United States Fish and Wildlife Service in 1960. The Service had carried out studies to discover whether fish, like warm-blooded animals, store insecticides in their tissues. The first samples were taken from forest areas in the West where there had been mass spraying of DDT for the control of the spruce budworm. As might have been expected, all of these fish contained DDT. The really significant findings were made when the investigators turned for comparison to a creek in a remote area about thirty miles from the nearest spraying for budworm control. This creek was upstream from the first and separated from it by a high waterfall. No local spraying was known to have occurred. Yet these fish, too, contained DDT. Had the chemical reached this remote creek by hidden underground streams? Or had it been airborne, drifting down as fallout on the surface of the creek? In still another comparative study, DDT was found in the tissues of fish from a hatchery where the water supply originated in a deep well. Again there was no record of local spraying. The only possible means of contamination seemed to be by means of ground-water.

In the entire water-pollution problem, there is probably nothing more disturbing than the threat of widespread con-tamination of ground-water. It is not possible to add pesticides to water anywhere without threatening the purity of water every-where. Seldom if ever does nature operate in closed and separate compartments, and she has not done so in distributing the earth's water supply. Rain, falling on the land, settles down through pores and cracks in soil and rock, penetrating deeper and deeper until eventually it reaches a zone where all the pores of the rock are filled with water, a dark, subsurface sea, rising under hills, sinking beneath valleys. This ground-water is always on the move, sometimes at a pace so slow that it travels no more than fifty feet a year, sometimes rapidly, by comparison, so that it moves nearly a tenth of a mile in a day. It travels by unseen waterways until here and there it comes to the surface as a spring, or perhaps it is tapped to feed a well. But mostly it contributes to streams and so to rivers. Except for what enters streams directly as rain or surface run-off, all the running water of the earth's surface was at one time ground-water. And so, in a very real and frightening sense, pollution of the ground-water is pollution of water everywhere.

It must have been by such a dark, underground sea that poisonous chemicals travelled from a manufacturing plant in Colorado to a farming district several miles away, there to poison wells, sicken humans and livestock, and damage crops – an extraordinary episode that may easily be only the first of many like it. Its history, in brief, is this. In 1943, the Rocky Mountain Arsenal of the Army Chemical Corps, located near Denver, began to manufacture war materials. Eight years later the facilities of the arsenal were leased to a private oil company for the production of insecticides. Even before the change of operations, however, mysterious reports had begun to come in. Farmers several miles from the plant began to report unexplained sickness among livestock; they complained of extensive crop damage. Foliage turned yellow, plants failed to mature, and many crops were killed outright. There were reports of human illness, thought by some to be related.

The irrigation waters on these farms were derived from shallow wells. When the well waters were examined (in a study in 1959, in which several state and federal agencies participated) they were found to contain an assortment of chemicals. Chlorides, chlorates, salts of phosphoric acid, fluorides, and arsenic had been discharged from the Rocky Mountain Arsenal into holding ponds during the years of its operation. Apparently the ground-water between the arsenal and the farms had become contaminated and it had taken seven to eight years for the wastes to travel underground a distance of about three miles from the holding ponds to the nearest farm. This seepage had continued to spread and had further contaminated an area of unknown extent. The investigators knew of no way to contain the contamination or halt its advance.

All this was bad enough, but the most mysterious and probably in the long run the most significant feature of the whole episode was the discovery of the weed killer 2,4-D in some of the wells and in the holding ponds of the arsenal. Certainly its presence was enough to account for the damage to crops irrigated with this water. But the mystery lay in the fact that no 2,4-D had been manufactured at the arsenal at any stage of its operations.

After long and careful study, the chemists at the plant concluded that the 2,4-D had been formed spontaneously in the open basins. It had been formed there from other substances discharged from the arsenal; in the presence of air, water, and sunlight, and quite without the intervention of human chemists, the holding ponds had become chemical laboratories for the production of a new chemical – a chemical fatally damaging to much of the plant life it touched.

And so the story of the Colorado farms and their damaged crops assumes a significance that transcends its local importance. What other parallels may there be, not only in Colorado but wherever chemical pollution finds its way into public waters? In lakes and streams everywhere, in the presence of catalysing air and sunlight, what dangerous substances may be born of parent chemicals labelled 'harmless'?

Indeed, one of the most alarming aspects of the chemical pollution of water is the fact that here – in river or lake or reservoir, or for that matter in the glass of water served at your dinner table – are mingled chemicals that no responsible chemist would think of combining in his laboratory. The possible interactions between these freely mixed chemicals are deeply disturbing to officials of the United States Public Health Service, who have expressed the fear that the production of harmful substances from comparatively innocuous chemicals may be taking place on quite a wide scale. The reactions may be between two or more chemicals, or between chemicals and the radioactive wastes that are being discharged into our rivers in ever-increasing volume. Under the impact of ionizing radiation some rearrangement of atoms could easily occur, changing the nature of the chemicals in a way that is not only unpredictable but beyond control.

It is, of course, not only the ground-waters that are becoming contaminated, but surface-moving waters as well – streams, rivers, irrigation waters. A disturbing example of the latter seems to be building up on the national wildlife refuges at Tule Lake and Lower Klamath, both in California. These refuges are part of a chain including also the refuge on Upper Klamath Lake just over the border in Oregon. All are linked, perhaps fatefully, by a shared water supply, and all are affected by the fact that they lie like small islands in a great sea of surrounding farmlands – land reclaimed by drainage and stream diversion from an original waterfowl paradise of marshland and open water.

These farmlands around the refuges are now irrigated by water from Upper Klamath Lake. The irrigation waters, re-collected from the fields they have served, are then pumped into Tule Lake and from there to Lower Klamath. All of the waters of the wildlife refuges established on these two bodies of water therefore represent the drainage of agricultural lands. It is important to remember this in connexion with recent happenings.

In the summer of 1960 the refuge staff picked up hundreds of dead and dying birds at Tule Lake and Lower Klamath. Most of them were fish-eating species – herons, pelicans, grebes, gulls. Upon analysis, they were found to contain insecticide residues identified as toxaphene, DDD, and DDE. Fish from the lakes were also found to contain insecticides; so did samples of plankton. The refuge manager believes that pesticide residues are now building up in the waters of these refuges, being conveyed there by return irrigation flow from heavily sprayed agricultural lands.

Such poisoning of waters set aside for conservation purposes could have consequences felt by every western duck hunter and by everyone to whom the sight and sound of drifting ribbons of waterfowl across an evening sky are precious. These particular refuges occupy critical positions in the conservation of western waterfowl. They lie at a point corresponding to the narrow neck of a funnel, into which all the migratory paths composing what is known as the Pacific Flyway converge. During the autumn migration they receive many millions of ducks and geese from nesting grounds extending from the shores of Bering Sea east to Hudson Bay – fully three-quarters of all the waterfowl that move south into the Pacific Coast states in autumn. In summer they provide nesting areas for waterfowl, especially for two endangered species, the redhead and the ruddy duck. If the lakes and pools of these refuges become seriously contaminated, the damage to the waterfowl populations of the Far West could be irreparable.

Water must also be thought of in terms of the chains of life it supports – from the small-as-dust green cells of the drifting plant plankton, through the minute water-fleas to the fishes that strain plankton from the water and are in turn eaten by other fishes or by birds, mink, raccoons – in an endless cyclic transfer of materials from life to life. We know that the necessary minerals in the water are so passed from link to link of the food chains. Can we suppose that poisons we introduce into water will not also enter into these cycles of nature?

The answer is to be found in the amazing history of Clear Lake, California. Clear Lake lies in mountainous country some ninety miles north of San Francisco and has long been popular with anglers. The name is inappropriate, for actually it is a rather turbid lake because of the soft black ooze that covers its shallow bottom. Unfortunately for the fishermen and the resort dwellers on its shores, its waters have provided an ideal habitat for a small gnat, *Chaoborus astictopus*. Although closely related to mosquitoes, the gnat is not a bloodsucker and probably does not feed at all as an adult. However, human beings who shared its habitat found it annoying because of its sheer numbers. Efforts were made to control it but they were largely fruitless until, in the late 1940s, the chlorinated hydrocarbon insecticides offered new weapons. The chemical chosen for a fresh attack was DDD, a close relative of DDT but apparently offering fewer threats to fish life.

The new control measures undertaken in 1949 were carefully planned and few people would have supposed any harm could result. The lake was surveyed, its volume determined, and the insecticide applied in such great dilution that for every part of chemical there would be seventy million parts of water. Control of the gnats was at first good, but by 1954 the treatment had to be repeated, this time at the rate of one part of insecticide in fifty million parts of water. The destruction of the gnats was thought to be virtually complete.

The following winter months brought the first intimation that other life was affected: the western grebes on the lake began to die, and soon more than a hundred of them were reported dead. At Clear Lake the western grebe is a breeding bird and also a winter visitant, attracted by the abundant fish of the lake. It is a bird of spectacular appearance and beguiling habits, building its floating nests in shallow lakes of western United States and Canada. It is called the 'swan grebe' with reason, for it glides with scarcely a ripple across the lake surface, the body riding low, white neck and shining black head held high. The newly hatched chick is clothed in soft grey down; in only a few hours it takes to the water and rides on the back of the father or mother, nestled under the parental wing coverts.

Following a third assault on the ever-resilient gnat population, in 1957, more grebes died. As had been true in 1954, no evidence of infectious disease could be discovered on examination of the dead birds. But when someone thought to analyse the fatty tissues of the grebes, they were found to be loaded with DDD in the extraordinary concentration of 1,600 parts per million.

The maximum concentration applied to the water was $\frac{1}{50}$ part per million. How could the chemical have built up to such prodigious levels in the grebes? These birds, of course, are fish-eaters. When the fish of Clear Lake also were analysed the picture began to take form – the poison being picked up by the smallest organisms, concentrated and passed on to the larger predators. Plankton organisms were found to contain about 5 parts per million of the insecticide (about twenty-five times the maximum concentration ever reached in the water itself); plant-eating fishes had built up accumulations ranging from 40 to 300 parts per million; carnivorous species had stored the most of all. One, a brown bullhead, had the astounding concentration of 2,500 parts per million. It was a house-that-Jack-built sequence, in which the large carnivores had eaten the smaller carnivores, that had eaten the herbivores, that had eaten the plankton, that had absorbed the poison from the water.

Even more extraordinary discoveries were made later. No trace of DDD could be found in the water shortly after the last application of the chemical. But the poison had not really left the lake; it had merely gone into the fabric of the life the lake supports. Twenty-three months after the chemical treatment had ceased, the plankton still contained as much as 5·3 parts per million. In that interval of nearly two years, successive crops of plankton had flowered and faded away, but the poison, although no longer present in the water, had somehow passed from generation to generation. And it lived on in the animal life of the lake as well. All fish, birds, and frogs examined a year after the chemical applications had ceased still contained DDD. The amount found in the flesh always exceeded by many times the original concentration in the water. Among these living carriers were fish that had hatched nine months after the last DDD application, grebes, and California gulls that had built up concentrations of more than 2,000 parts per million. Meanwhile, the nesting colonies of the grebes dwindled – from more than 1,000 pairs before the first insecticide treatment to about thirty pairs in 1960. And even the thirty seem to have nested in vain, for no young grebes have been observed on the lake since the last DDD application.

This whole chain of poisoning, then, seems to rest on a base of minute plants which must have been the original concentrators. But what of the opposite end of the food chain – the human being who, in probable ignorance of all this sequence of events, has rigged his fishing tackle, caught a string of fish from the waters of Clear Lake, and taken them home to fry for his supper? What could a heavy dose of DDD, or perhaps repeated doses, do to him?

Although the California Department of Public Health professed to see no hazard, nevertheless in 1959 it required that the use of DDD in the lake be stopped. In view of scientific evidence of the vast biological potency of this chemical, the action seems a minimum safety measure. The physiological effect of DDD is probably unique among insecticides, for it destroys part of the adrenal gland – the cells of the outer layer known as the adrenal cortex, which secretes the hormone cortin. This destructive effect, known since 1948, was at first believed to be confined to dogs, because it was not revealed in such experimental animals as monkeys, rats, or rabbits. It seemed suggestive, however, that DDD produced in dogs a condition very similar to that occurring in man in the presence of Addison's disease. Recent medical research has revealed that DDD does strongly suppress the function of the human adrenal cortex. Its cell-destroying capacity is now clinically utilized in the treatment of a rare type of cancer which develops in the adrenal gland.

The Clear Lake situation brings up a question that the public needs to face: Is it wise or desirable to use substances with such strong effect on physiological processes for the control of insects, especially when the control measures involve introducing the chemical directly into a body of water? The fact that the insecticide was applied in very low concentrations is meaningless, as its explosive progress through the natural food chain in the lake demonstrates. Yet Clear Lake is typical of a large and growing number of situations where solution of an obvious and often trivial problem creates a far more serious but conveniently less tangible one. Here the problem was resolved in favour of those annoyed by gnats, and at the expense of an unstated, and probably not even clearly understood, risk to all who took food or water from the lake.

It is an extraordinary fact that the deliberate introduction of poisons into a reservoir is becoming a fairly common practice. The purpose is usually to promote recreational uses, even though the water must then be treated at some expense to make it fit for its intended use as drinking water. When sportsmen of an area want to 'improve' fishing in a reservoir, they prevail on authorities to dump quantities of poison into it to kill the undesired fish, which are then replaced with hatchery fish more suited to the sportsmen's taste. The procedure has a strange, Alice-in-Wonderland quality. The reservoir was created as a public water supply, yet the community, probably unconsulted about the sportsmen's project, is forced either to drink water containing poisonous residues or to pay out tax money for treatment of the water to remove the poisons – treatments that are by no means foolproof.

As ground and surface waters are contaminated with pesticides and other chemicals, there is danger that not only poisonous but also cancer-producing substances are being introduced into public water supplies. Dr W. C. Hueper of the National Cancer Institute has warned that

the danger of cancer hazards from the consumption of contaminated drinking water will grow considerably within the foreseeable future.

And indeed a study made in Holland in the early 1950s provides support for the view that polluted waterways may carry a cancer hazard. Cities receiving their drinking water from rivers had a higher death-rate from cancer than did those whose water came from sources presumably less susceptible to pollution such as wells. Arsenic, the environmental substance most clearly established as causing cancer in man, is involved in two historic cases in which polluted water supplies caused widespread occurrence of cancer. In one case the arsenic came from the slag heaps of mining operations, in the other from rock with a high natural content of arsenic. These conditions may easily be duplicated as a result of heavy applications of arsenical insecticides. The soil in such areas becomes poisoned. Rains then carry part of the arsenic into streams, rivers, and reservoirs, as well as into the vast subterranean seas of ground-water.

Here again we are reminded that in nature nothing exists alone. To understand more clearly how the pollution of our world is happening, we must now look at another of the earth's basic resources, the soil.

Chapter 5

Realms of the Soil

The thin layer of soil that forms a patchy covering over the continents controls our own existence and that of every other animal of the land. Without soil, land plants as we know them could not grow, and without plants no animals could survive.

Yet if our agriculture-based life depends on the soil, it is equally true that soil depends on life, its very origins and the maintenance of its true nature being intimately related to living plants and animals. For soil is in part a creation of life, born of a marvellous interaction of life and non-life long aeons ago. The parent materials were gathered together as volcanoes poured them out in fiery streams, as waters running over the bare rocks of the continents wore away even the hardest granite, and as the chisels of frost and ice split and shattered the rocks. Then living things began to work their creative magic and little by little these inert materials became soil. Lichens, the rocks' first covering, aided the process of disintegration by their acid secretions and made a lodging place for other life. Mosses took hold in the little pockets of simple soil – soil formed by crumbling bits of lichen, by the husks of minute insect life, by the debris of a fauna beginning its emergence from the sea.

Life not only formed the soil, but other living things of incredible abundance and diversity now exist within it; if this were not so the soil would be a dead and sterile thing. By their presence and by their activities the myriad organisms of the soil make it capable of supporting the earth's green mantle.

The soil exists in a state of constant change, taking part in cycles that have no beginning and no end. New materials are constantly being contributed as rocks disintegrate, as organic matter decays, and as nitrogen and other gases are brought down in rain from the skies. At the same time other materials are being taken away, borrowed for temporary use by living creatures.

Subtle and vastly important chemical changes are constantly in progress, converting elements derived from air and water into forms suitable for use by plants. In all these changes living organisms are active agents.

There are few studies more fascinating, and at the same time more neglected, than those of the teeming populations that exist in the dark realms of the soil. We know too little of the threads that bind the soil organisms to each other and to their world, and to the world above.

Perhaps the most essential organisms in the soil are the smallest – the invisible hosts of bacteria and of threadlike fungi. Statistics of their abundance take us at once into astronomical figures. A teaspoonful of topsoil may contain billions of bacteria. In spite of their minute size, the total weight of this host of bacteria in the top foot of a single acre of fertile soil may be as much as a thousand pounds. Ray fungi, growing in long thread-like filaments, are somewhat less numerous than the bacteria, yet because they are larger their total weight in a given amount of soil may be about the same. With small green cells called algae, these make up the microscopic plant life of the soil.

Bacteria, fungi, and algae are the principal agents of decay, reducing plant and animal residues to their component minerals. The vast cyclic movements of chemical elements such as carbon and nitrogen through soil and air and living tissue could not proceed without these microplants. Without the nitrogen-fixing bacteria, for example, plants would starve for want of nitrogen, though surrounded by a sea of nitrogen-containing air. Other organisms form carbon dioxide, which, as carbonic acid, aids in dissolving rock. Still other soil microbes perform various oxidations and reductions by which minerals such as iron, manganese, and sulphur are transformed and made available to plants.

Also present in prodigious numbers are microscopic mites and primitive wingless insects called springtails. Despite their small size they play an important part in breaking down the residues of plants, aiding in the slow conversion of the litter of the forest floor to soil. The specialization of some of these minute creatures for their task is almost incredible. Several species of mites, for example, can begin life only within the fallen needles of a spruce tree. Sheltered here, they digest out the inner tissues of the needle. When the mites have completed their development only the outer layer of cells remains. The truly staggering task of dealing with the tremendous amount of plant material in the annual leaf fall belongs to some of the small insects of the soil and the forest floor. They macerate and digest the leaves, and aid in mixing the decomposed matter with the surface soil.

Besides all this horde of minute but ceaselessly toiling creatures there are of course many larger forms, for soil life runs the gamut from bacteria to mammals. Some are permanent residents of the dark subsurface layers; some hibernate or spend definite parts of their life cycles in underground chambers; some freely come and go between their burrows and the upper world. In general the effect of all this habitation of the soil is to aerate it and improve both its drainage and the penetration of water throughout the layers of plant growth.

Of all the larger inhabitants of the soil, probably none is more important than the earthworm. Over three-quarters of a century ago, Charles Darwin published a book entitled *The Formation of Vegetable Mould, through the Action of Worms, with Observations on Their Habits.* In it he gave the world its first understanding of the fundamental role of earthworms as geologic agents for the transport of soil – a picture of surface rocks being gradually covered by fine soil brought up from below by the worms, in annual amounts running to many tons to the acre in most favourable areas. At the same time, quantities of organic matter contained in leaves and grass (as much as twenty pounds to the square yard in six months) are drawn down into the burrows and incorporated in soil. Darwin's calculations showed that the toil of earthworms might add a layer of soil an inch to an inch and a half thick in a ten-year period. And this is by no means all they do: their burrows aerate the soil, keep it well drained, and aid the penetration of plant roots. The presence of earthworms increases the nitrifying powers of the soil bacteria and decreases putrefaction of the soil. Organic matter is broken down as it passes through the digestive tracts of the worms and the soil is enriched by their excretory products.

This soil community, then, consists of a web of interwoven lives, each in some way related to the others – the living creatures depending on the soil, but the soil in turn a vital element of the earth only so long as this community within it flourishes.

The problem that concerns us here is one that has received little consideration: What happens to these incredibly numerous and vitally necessary inhabitants of the soil when poisonous chemicals are carried down into their world, either introduced directly as soil 'sterilants' or borne on the rain that has picked up a lethal contamination as it filters through the leaf canopy of forest and orchard and cropland? Is it reasonable to suppose that we can apply a broad-spectrum insecticide to kill the burrowing larval stages of a crop-destroying insect, for example, without also killing the 'good' insects whose function may be the essential one of breaking down organic matter? Or can we use a non-specific fungicide without also killing the fungi that inhabit the roots of many trees in a beneficial association that aids the tree in extracting nutrients from the soil?

The plain truth is that this critically important subject of the ecology of the soil has been largely neglected even by scientists and almost completely ignored by control men. Chemical control of insects seems to have proceeded on the assumption that the soil could and would sustain any amount of insult via the introduction of poisons without striking back. The very nature of the world of the soil has been largely ignored.

From the few studies that have been made, a picture of the impact of pesticides on the soil is slowly emerging. It is not surprising that the studies are not always in agreement, for soil types vary so enormously that what causes damage in one may be innocuous in another. Light sandy soils suffer far more heavily than humus types. Combinations of chemicals seem to do more harm than separate applications. Despite the varying results, enough solid evidence of harm is accumulating to cause apprehension on the part of many scientists.

Under some conditions, the chemical conversions and transformations that lie at the very heart of the living world are affected. Nitrification, which makes atmospheric nitrogen available to plants, is an example. The herbicide 2,4-D causes a temporary interruption of nitrification. In recent experiments in Florida, lindane, heptachlor, and BHC (benzene hexachloride) reduced nitrification after only two weeks in soil; BHC and

DDT had significantly detrimental effects a year after treatment. In other experiments BHC, aldrin, lindane, heptachlor, and DDD all prevented nitrogen-fixing bacteria from forming the necessary root nodules on leguminous plants. A curious but beneficial relation between fungi and the roots of higher plants is seriously disrupted.

Sometimes the problem is one of upsetting that delicate balance of populations by which nature accomplishes far-reaching aims. Explosive increases in some kinds of soil organisms have occurred when others have been reduced by insecticides, disturbing the relation of predator to prey. Such changes could easily alter the metabolic activity of the soil and affect its productivity. They could also mean that potentially harmful organisms, formerly held in check, could escape from their natural controls and rise to pest status.

One of the most important things to remember about insecticides in soil is their long persistence, measured not in months but in years. Aldrin has been recovered after four years, both as traces and more abundantly as converted to dieldrin. Enough toxaphene remains in sandy soil ten years after its application to kill termites. Benzene hexachloride persists at least eleven years; heptachlor or a more toxic derived chemical, at least nine. Chlordane has been recovered twelve years after its application, in the amount of 15 per cent of the original quantity.

Seemingly moderate applications of insecticides over a period of years may build up fantastic quantities in soil. Since the chlorinated hydrocarbons are persistent and long-lasting, each application is merely added to the quantity remaining from the previous one. The old legend that 'a pound of DDT to the acre is harmless' means nothing if spraying is repeated. Potato soils have been found to contain up to 15 pounds of DDT per acre, corn soils up to 19. A cranberry bog under study contained 34·5 pounds to the acre. Soils from apple orchards seem to reach the peak of contamination, with DDT accumulating at a rate that almost keeps pace with its rate of annual application. Even in a single season, with orchards sprayed four or more times, DDT residues may build up to peaks of 30 to 50 pounds. With repeated spraying over the years the range between trees is from 26 to 60 pounds to the acre; under trees, up to 113 pounds.

Arsenic provides a classic case of the virtually permanent poisoning of the soil. Although arsenic as a spray on growing tobacco has been largely replaced by the synthetic organic insecticides since the mid '40s, *the arsenic content of cigarettes made from American-grown tobacco increased more than 300 per cent* between the years 1932 and 1952. Later studies have revealed increases of as much as 600 per cent. Dr Henry S. Satterlee, an authority on arsenic toxicology, says that although organic insecticides have been largely substituted for arsenic, the tobacco plants continue to pick up the old poison, for the soils of tobacco plantations are now thoroughly impregnated with residues of a heavy and relatively insoluble poison, arsenate of lead. This will continue to release arsenic in soluble form. The soil of a large proportion of the land planted to tobacco has been subjected to 'cumulative and well-nigh permanent poisoning', according to Dr Satterlee. Tobacco grown in the eastern Mediterranean countries where arsenical insecticides are not used has shown no such increase in arsenic content.

We are therefore confronted with a second problem. We must not only be concerned with what is happening to the soil; we must wonder to what extent insecticides are absorbed from contaminated soils and introduced into plant tissues. Much depends on the type of soil, the crop, and the nature and concentration of the insecticide. Soil high in organic matter releases smaller quantities of poisons than others. Carrots absorb more insecticide than any other crop studied; if the chemical used happens to be lindane, carrots actually accumulate higher concentrations than are present in the soil. In the future it may become necessary to analyse soils for insecticides before planting certain food crops. Otherwise even unsprayed crops may take up enough insecticide merely from the soil to render them unfit for market.

This very sort of contamination has created endless problems for at least one leading manufacturer of baby foods who has been unwilling to buy any fruits or vegetables on which toxic insecticides have been used. The chemical that caused him the most trouble was benzene hexachloride (BHC), which is taken up by the roots and tubers of plants, advertising its presence by a musty taste and odour. Sweet potatoes grown on California fields where

BHC had been used two years earlier contained residues and had to be rejected. In one year, in which the firm had contracted in South Carolina for its total requirements of sweet potatoes, so large a proportion of the acreage was found to be contaminated that the company was forced to buy in the open market at a considerable financial loss. Over the years a variety of fruits and vegetables, grown in various states, have had to be rejected. The most stubborn problems were concerned with peanuts. In the southern states peanuts are usually grown in rotation with cotton, on which BHC is extensively used. Peanuts grown later in this soil pick up considerable amounts of the insecticide. Actually, only a trace is enough to incorporate the telltale musty odour and taste. The chemical penetrates the nuts and cannot be removed. Processing, far from removing the mustiness, sometimes accentuates it. The only course open to a manufacturer determined to exclude BHC residues is to reject all produce treated with the chemical or grown on soils contaminated with it.

Sometimes the menace is to the crop itself – a menace that remains as long as the insecticide contamination is in the soil. Some insecticides affect sensitive plants such as beans, wheat, barley, or rye, retarding root development or depressing growth of seedlings. The experience of the hop growers in Washington and Idaho is an example. During the spring of 1955 many of these growers undertook a large-scale programme to control the strawberry root weevil, whose larvae had become abundant on the roots of the hops. On the advice of agricultural experts and insecticide manufacturers, they chose heptachlor as the control agent. Within a year after the heptachlor was applied, the vines in the treated yards were wilting and dying. In the untreated fields there was no trouble; the damage stopped at the border between treated and untreated fields. The hills were replanted at great expense, but in another year the new roots, too, were found to be dead. Four years later the soil still contained heptachlor, and scientists were unable to predict how long it would remain poisonous, or to recommend any procedure for correcting the condition. The federal Department of Agriculture, which as late as March 1959 found itself in the anomalous position of declaring heptachlor to be acceptable for use on hops in the form of a soil treatment, belatedly withdrew its registration for such use. Meanwhile, the hop growers sought what redress they could in the courts.

As applications of pesticides continue and the virtually in-destructible residues continue to build up in the soil, it is almost certain that we are heading for trouble. This was the consensus of a group of specialists who met at Syracuse University in 1960 to discuss the ecology of the soil. These men summed up the hazards of using 'such potent and little understood tools' as chemicals and radiation:

A few false moves on the part of man may result in destruction of soil productivity and the arthropods may well take over.

Chapter 6

Earth's Green Mantle

Water, soil, and the earth's green mantle of plants make up the world that supports the animal life of the earth. Although modern man seldom remembers the fact, he could not exist without the plants that harness the sun's energy and manufacture the basic foodstuffs he depends upon for life. Our attitude towards plants is a singularly narrow one. If we see any immediate utility in a plant we foster it. If for any reason we find its presence undesirable or merely a matter of indifference, we may condemn it to destruction forthwith. Besides the various plants that are poisonous to man or his livestock, or crowd out food plants, many are marked for destruction merely because, according to our narrow view, they happen to be in the wrong place at the wrong time. Many others are destroyed merely because they happen to be associates of the unwanted plants.

The earth's vegetation is part of a web of life in which there are intimate and essential relations between plants and the earth, between plants and other plants, between plants and animals. Sometimes we have no choice but to disturb these relationships, but we should do so thoughtfully, with full awareness that what we do may have consequences remote in time and place. But no such humility marks the booming 'weed killer' business of the present day, in which soaring sales and expanding uses mark the production of plant-killing chemicals.

One of the most tragic examples of our unthinking bludgeoning of the landscape is to be seen in the sagebrush lands of the West, where a vast campaign is on to destroy the sage and to substitute grasslands. If ever an enterprise needed to be illuminated with a sense of the history and meaning of the landscape, it is this. For here the natural landscape is eloquent of the interplay of forces that have created it. It is spread before us like the pages of an open book in which we can read why the land is what it is, and why we should preserve its integrity. But the pages lie unread.

The land of the sage is the land of the high western plains and the lower slopes of the mountains that rise above them, a land born of the great uplift of the Rocky Mountain system many millions of years ago. It is a place of harsh extremes of climate: of long winters when blizzards drive down from the mountains and snow lies deep on the plains, of summers whose heat is relieved by only scanty rains, with drought biting deep into the soil, and drying winds stealing moisture from leaf and stem.

As the landscape evolved, there must have been a long period of trial and error in which plants attempted the colonization of this high and windswept land. One after another must have failed. At last one group of plants evolved which combined all the qualities needed to survive. The sage – low-growing and shrubby – could hold its place on the mountain slopes and on the plains, and within its small grey leaves it could hold moisture enough to defy the thieving winds. It was no accident, but rather the result of long ages of experimentation by nature, that the great plains of the West became the land of the sage.

Along with the plants, animal life, too, was evolving in harmony with the searching requirements of the land. In time there were two as perfectly adjusted to their habitat as the sage. One was a mammal, the fleet and graceful pronghorn antelope. The other was a bird, the sage grouse – the 'cock of the plains' of Lewis and Clark.

The sage and the grouse seem made for each other. The original range of the bird coincided with the range of the sage, and as the sagelands have been reduced, so the populations of grouse have dwindled. The sage is all things to these birds of the plains. The low sage of the foothill ranges shelters their nests and their young; the denser growths are loafing and roosting areas; at all times the sage provides the staple food of the grouse. Yet it is a two-way relationship. The spectacular courtship displays of the cocks help loosen the soil beneath and around the sage, aiding invasion by grasses which grow in the shelter of sagebrush.

The antelope, too, have adjusted their lives to the sage. They are primarily animals of the plains, and in winter when the first snows come those that have summered in the mountains move down to the lower elevations. There the sage provides the food that tides them over the winter. Where all other plants have shed their leaves, the sage remains evergreen, the grey-green leaves – bitter, aromatic, rich in protein, fats, and needed minerals – clinging to the stems of the dense and shrubby plants. Though the snow piles up, the tops of the sage remain exposed, or can be reached by the sharp, pawing hoofs of the antelope. Then grouse feed on them too, finding them on bare and windswept ledges or following the antelope to feed where they have scratched away the snow.

And other life looks to the sage. Mule deer often feed on it. Sage may mean survival for winter-grazing livestock. Sheep graze many winter ranges where the big sagebrush forms almost pure stands. For half the year it is their principal forage, a plant of higher energy value than even alfalfa hay.

The bitter upland plains, the purple wastes of sage, the wild, swift antelope, and the grouse are then a natural system in perfect balance. Are? The verb must be changed – at least in those already vast and growing areas where man is attempting to improve on nature's way. In the name of progress the land management agencies have set about to satisfy the insatiable demands of the cattlemen for more grazing land. By this they mean grassland – grass without sage. So in a land which nature found suited to grass growing mixed with and under the shelter of sage, it is now proposed to eliminate the sage and create unbroken grassland. Few seem to have asked whether grasslands are a stable and desirable goal in this region. Certainly nature's own answer was otherwise. The annual precipitation in this land where the rains seldom fall is not enough to support good sod-forming grass; it favours rather the perennial bunch-grass that grows in the shelter of the sage.

Yet the programme of sage eradication has been under way for a number of years. Several government agencies are active in it; industry has joined with enthusiasm to promote and encourage an enterprise which creates expanded markets not only for grass seed but for a large assortment of machines for cutting and ploughing and seeding. The newest addition to the weapons is the use of chemical sprays. Now millions of acres of sagebrush lands are sprayed each year.

What are the results? The eventual effects of eliminating sage and seeding with grass are largely conjectural. Men of long experience with the ways of the land say that in this country there is better growth of grass between and under the sage than can possibly be had in pure stands, once the moisture-holding sage is gone.

But even if the programme succeeds in its immediate objective, it is clear that the whole closely-knit fabric of life has been ripped apart. The antelope and the grouse will disappear along with the sage. The deer will suffer, too, and the land will be poorer for the destruction of the wild things that belong to it. Even the livestock which are the intended beneficiaries will suffer; no amount of lush green grass in summer can help the sheep starving in the winter storms for lack of the sage and bitterbrush and other wild vegetation of the plains.

These are the first and obvious effects. The second is of a kind that is always associated with the shot-gun approach to nature: the spraying also eliminates a great many plants that were not its intended target. Justice William O. Douglas, in his recent book *My Wilderness: East to Katahdin*, has told of an appalling example of ecological destruction wrought by the United States Forest Service in the Bridger National Forest in Wyoming. Some 10,000 acres of sagelands were sprayed by the Service, yielding to pressure of cattlemen for more grasslands. The sage was killed, as intended. But so was the green, life-giving ribbon of willows that traced its way across these plains, following the meandering streams. Moose had lived in these willow thickets, for willow is to the moose what sage is to the antelope. Beaver had lived there, too, feeding on the willows, felling them and making a strong dam across the tiny stream. Through the labour of the beavers, a lake backed up. Trout in the mountain streams seldom were more than six inches long; in the lake they thrived so prodigiously that many grew to five pounds. Waterfowl were attracted to the lake, also. Merely because of the presence of the willows and the beavers that depended on them, the region was an attractive recreational area with excellent fishing and hunting.

But with the 'improvement' instituted by the Forest Service, the willows went the way of the sagebrush, killed by the same impartial spray. When Justice Douglas visited the area in 1959, the year of the spraying, he was shocked to see the shrivelled and dying willows – the 'vast, incredible damage'. What would become of the moose? Of the beavers and the little world they had constructed? A year later he returned to read the answers in the devastated landscape. The moose were gone and so were the beaver. Their principal dam had gone out for want of attention by its skilled architects, and the lake had drained away. None of the large trout were left. None could live in the tiny creek that remained, threading its way through a bare, hot land where no shade remained. The living world was shattered.

Besides the more than four million acres of rangelands sprayed each year, tremendous areas of other types of land are also potential or actual recipients of chemical treatments for weed control. For example, an area larger than all of New England – some 50 million acres – is under management by utility corporations and much of it is routinely treated for 'brush control'. In the South-west an estimated 75 million acres of mesquite lands require management by some means, and chemical spraying is the method most actively pushed. An unknown but very large acreage of timber-producing lands is now aerially sprayed in order to 'weed out' the hardwoods from the more spray-resistant conifers. Treatment of agricultural lands with herbicides doubled in the decade following 1949, totalling 53 million acres in 1959. And the combined acreage of private lawns, parks, and golf courses now being treated must reach an astronomical figure.

The chemical weed killers are a bright new toy. They work in a spectacular way; they give a giddy sense of power over nature to those who wield them, and as for the long-range and less obvious effects – these are easily brushed aside as the baseless imaginings of pessimists. The 'agricultural engineers' speak blithely of 'chemical ploughing' in a world that is urged to beat its ploughshares into spray guns. The town fathers of a thousand communities lend willing ears to the chemical salesmen and the eager contractors who will rid the roadsides of 'brush' – for a price. It is cheaper than mowing, is the cry. So, perhaps, it appears in the neat rows of figures in the official books; but were the true costs entered, the costs not only in dollars but in the many equally valid debits we shall presently consider, the wholesale broadcasting of chemicals would be seen to be more costly in dollars as well as infinitely damaging to the long-range health of the landscape and to all the varied interests that depend on it.

Take, for instance, the commodity prized by every chamber of commerce throughout the land – the good will of vacationing tourists. There is a steadily growing chorus of outraged protest about the disfigurement of once-beautiful roadsides by chemical sprays, which substitute a sere expanse of brown, withered vegetation for the beauty of fern and wildflower, of native shrubs adorned with blossom or berry.

We are making a dirty, brown, dying-looking mess along the sides of our roads [a New England woman wrote angrily to her newspaper]. This is not what the tourists expect, with all the money we are spending advertising the beautiful scenery.

In the summer of 1960 conservationists from many states converged on a peaceful Maine island to witness its presentation to the National Audubon Society by its owner, Millicent Todd Bingham. The focus that day was on the preservation of the natural landscape and of the intricate web of life whose interwoven strands lead from microbes to man. But in the background of all the conversations among the visitors to the island was indignation at the despoiling of the roads they had travelled. Once it had been a joy to follow those roads through the evergreen forests, roads lined with bayberry and sweet fern, alder and huckleberry. Now all was brown desolation. One of the conservationists wrote of that August pilgrimage to a Maine island:

I returned ... angry at the desecration of the Maine roadsides . Where, in previous years, the highways were bordered with wildflowers and attractive shrubs, there were only the scars of dead vegetation for mile after mile.... As an economic proposition, can Maine afford the loss of tourist good will that such sights induce?

Maine roadsides are merely one example, though a particularly sad one for those of us who have a deep love for the beauty of that state, of the senseless destruction that is going on in the name of roadside brush control throughout the nation.

Botanists at the Connecticut Arboretum declare that the elimination of beautiful native shrubs and wildflowers has reached the proportions of a 'roadside crisis'. Azaleas, mountain laurel, blueberries, huckleberries, viburnums, dogwood, bayberry, sweet fern, low shadbush, winterberry, chokecherry, and wild plum are dying before the chemical barrage. So are the daisies, black-eyed Susans, Queen Anne's lace, goldenrods, and autumn asters which lend grace and beauty to the landscape.

The spraying is not only improperly planned but studded with abuses such as these. In a southern New England town one contractor finished his work with some chemical remaining in his tank. He discharged this along woodland roadsides where no spraying had been authorized. As a result the community lost the blue and golden beauty of its autumn roads, where asters and goldenrod would have made a display worth travelling far to see. In another New England community a contractor changed the state specifications for town spraying without the knowledge of the highway department and sprayed roadside vegetation to a height of eight feet instead of the specified maximum of four feet, leaving a broad, disfiguring, brown swath. In a Massachusetts community the town officials purchased a weed killer from a zealous chemical salesman, unaware that it contained arsenic. One result of the subsequent roadside spraying was the death of a dozen cows from arsenic poisoning.

Trees within the Connecticut Arboretum Natural Area were seriously injured when the town of Waterford sprayed the roadsides with chemical weed killers in 1957. Even large trees not directly sprayed were affected. The leaves of the oaks began to curl and turn brown, although it was the season for spring growth. Then new shoots began to be put forth and grew with abnormal rapidity, giving a weeping appearance to the trees. Two seasons later, large branches on these trees had died, others were without leaves, and the deformed, weeping effect of whole trees persisted.

I know well a stretch of road where nature's own landscaping has provided a border of alder, viburnum, sweet fern, and juniper with seasonally changing accents of bright flowers, or of fruits hanging in jewelled clusters in the autumn. The road had no heavy load of traffic to support; there were few sharp curves or intersections where brush could obstruct the driver's vision. But the sprayers took over and the miles along that road became something to be traversed quickly, a sight to be endured with one's mind closed to thoughts of the sterile and hideous world we are letting our technicians make. But here and there authority had somehow faltered and by an unaccountable oversight there were oases of beauty in the midst of austere and regimented control – oases that made the desecration of the greater part of the road the more unbearable. In such places my spirit lifted to the sight of the drifts of white clover or the clouds of purple vetch with here and there the flaming cup of a wood lily.

Such plants are 'weeds' only to those who make a business of selling and applying chemicals. In a volume of *Proceedings* of one of the weed-control conferences that are now regular institutions, I once read an extraordinary statement of a weed killer's philosophy. The author defended the killing of good plants 'simply because they are in bad company'. Those who complain about killing wildflowers along roadsides reminded him, he said, of antivivisectionists 'to whom, if one were to judge by their actions, the life of a stray dog is more sacred than the lives of children'.

To the author of this paper, many of us would unquestionably be suspect, convicted of some deep perversion of character because we prefer the sight of the vetch and the clover and the wood lily in all their delicate and transient beauty to that of roadsides scorched as by fire, the shrubs brown and brittle, the bracken that once lifted high its proud lacework now withered and drooping. We would seem deplorably weak that we can tolerate the sight of such 'weeds', that we do not rejoice in their eradication, that we are not filled with exultation that man has once more triumphed over miscreant nature.

Justice Douglas tells of attending a meeting of federal field men who were discussing protests by citizens against plans for the spraying of sagebrush that I mentioned earlier in this chapter. These men considered it hilariously funny that an old lady had opposed the plan because the wildflowers would be destroyed.

Yet, was not her right to search out a banded cup or a tiger lily as inalienable as the right of stockmen to search out grass or of a lumberman to claim a tree? [asks this humane and perceptive jurist.] The aesthetic values of the wilderness are as much our inheritance as the veins of copper and gold in our hills and the forests in our mountains.

There is, of course, more to the wish to preserve our roadside vegetation than even such aesthetic considerations. In the economy of nature the natural vegetation has its essential place. Hedgerows along country roads and bordering fields provide food, cover, and nesting areas for birds and homes for many small animals. Of some seventy species of shrubs and vines that are typical roadside species in the eastern states alone, about sixty-five are important to wildlife as food.

Such vegetation is also the habitat of wild bees and other pollinating insects. Man is more dependent on these wild pollinators than he usually realizes. Even the farmer himself seldom understands the value of wild bees and often participates in the very measures that rob him of their services. Some agricultural crops and many wild plants are partly or wholly dependent on the services of the native pollinating insects. Several hundred species of wild bees take part in the pollination of cultivated crops – 100 species visiting the flowers of alfalfa alone. Without insect pollination, most of the soil-holding and soil-enriching plants of uncultivated areas would die out, with far-reaching consequences to the ecology of the whole region. Many herbs, shrubs, and trees of forests and range depend on native insects for their reproduction; without these plants many wild animals and range stock would find little food. Now clean cultivation and the chemical destruction of hedgerows and weeds are eliminating the last sanctuaries of these pollinating insects and breaking the threads that bind life to life.

These insects, so essential to our agriculture and indeed to our landscape as we know it, deserve something better from us than the senseless destruction of their habitat. Honeybees and wild bees depend heavily on such 'weeds' as goldenrod, mustard, and dandelions for pollen that serves as the food of their young. Vetch furnishes essential spring forage for bees before the alfalfa is in bloom, tiding them over this early season so that they are ready to pollinate the alfalfa. In the autumn they depend on goldenrod at a season when no other food is available, to stock up for the winter. By the precise and delicate timing that is nature's own, the emergence of one species of wild bees takes place on the very day of the opening of the willow blossoms. There is no dearth of men who understand these things, but these are not the men who order the wholesale drenching of the landscape with chemicals.

And where are the men who supposedly understand the value of proper habitat for the preservation of wildlife? Too many of them are to be found defending herbicides as 'harmless' to wildlife because they are thought to be less toxic than insecticides. Therefore, it is said, no harm is done. But as the herbicides rain down on forest and field, on marsh and rangeland, they are bringing about marked changes and even permanent destruction of wildlife habitat. To destroy the homes and the food of wildlife is perhaps worse in the long run than direct killing.

The irony of this all-out chemical assault on roadsides and utility rights-of-way is twofold. It is perpetuating the problem it seeks to correct, for as experience has clearly shown, the blanket application of herbicides does not permanently control roadside 'brush' and the spraying has to be repeated year after year. And as a further irony, we persist in doing this despite the fact that a perfectly sound method of *selective* spraying is known, which can achieve long-term vegetational control and eliminate repeated spraying in most types of vegetation.

The object of brush control along roads and rights-of-way is not to sweep the land clear of everything but grass; it is, rather, to eliminate plants ultimately tall enough to present an obstruction to drivers' vision or interference with wires on rights-of-way. This means, in general, trees. Most shrubs are low enough to present no hazard; so, certainly, are ferns and wildflowers.

Selective spraying was developed by Dr Frank Egler during a period of years at the American Museum of Natural History as director of a Committee for Brush Control Recommendations for Rights-of-Way. It took advantage of the inherent stability of nature, building on the fact that most communities of shrubs are strongly resistant to invasion by trees. By comparison, grasslands are easily invaded by tree seedlings. The object of selective spraying is not to produce grass on roadsides and rights-of-way but to eliminate the tall woody plants by direct treatment and to preserve all other vegetation. One treatment may be sufficient, with a possible follow-up for extremely resistant species; thereafter the shrubs assert control and the trees do not return. The best and cheapest controls for vegetation are not chemicals but other plants.

The method has been tested in research areas scattered throughout the eastern United States. Results show that, once properly treated, an area becomes stabilized, *requiring no respraying for at least twenty years*. The spraying can often be done by men on foot, using knapsack sprayers, and having complete control over their material. Sometimes compressor pumps and material can be mounted on truck chassis, but there is no blanket spraying. Treatment is directed only to trees and exceptionally tall shrubs that must be eliminated. The integrity of the environment is thereby preserved, the enormous value of the wildlife habitat remains intact, and the beauty of shrub and fern and wildflower has not been sacrificed.

Here and there the method of vegetation management by selective spraying has been adopted. For the most part, entrenched custom dies hard and blanket spraying continues to thrive, to exact its heavy annual costs from the taxpayer, and to inflict its damage on the ecological web of life. It thrives, surely, only because the facts are not known. When taxpayers understand that the bill for spraying the town roads should come due only once a generation instead of once a year, they will surely rise up and demand a change of method.

Among the many advantages of selective spraying is the fact that it minimizes the amount of chemical applied to the landscape. There is no broadcasting of material but, rather, concentrated application to the base of the trees. The potential harm to wildlife is therefore kept to a minimum.

The most widely used herbicides are 2,4-D, 2,4,5-T, and related compounds. Whether or not these are actually toxic is a matter of controversy. People spraying their lawns with 2,4-D and becoming wet with spray have occasionally developed severe neuritis and even paralysis. Although such incidents are apparently uncommon, medical authorities advise caution in use of such compounds. Other hazards, more obscure, may also attend the use of 2,4-D. It has been shown experimentally to disturb the basic physiological process of respiration in the cell, and to imitate X-rays in damaging the chromosomes. Some very recent work indicates that reproduction of birds may be adversely affected by these and certain other herbicides at levels far below those that cause death.

Apart from any directly toxic effects, curious indirect results follow the use of certain herbicides. It has been found that animals, both wild herbivores and livestock, are sometimes strangely attracted to a plant that has been sprayed, even though it is not one of their natural foods. If a highly poisonous herbicide such as arsenic has been used, this intense desire to reach the wilting vegetation inevitably has disastrous results. Fatal results may follow, also, from less toxic herbicides if the plant itself happens to be poisonous or perhaps to possess thorns or burs. Poisonous range weeds, for example, have suddenly become attractive to livestock after spraying, and the animals have died from indulging this unnatural appetite. The literature of veterinary medicine abounds in similar examples, swine eating sprayed cockleburs with consequent severe illness, lambs eating sprayed thistles, bees poisoned by pasturing on mustard sprayed after it came into bloom. Wild cherry, the leaves of which are highly poisonous, has exerted a fatal attraction for cattle once its foliage has been sprayed with 2,4-D. Apparently the wilting that follows spraying (or cutting) makes the plant attractive. Ragwort has provided other examples. Livestock ordinarily avoid this plant unless forced to turn to it in late winter and early spring by lack of other forage. However, the animals eagerly feed on it after its foliage has been sprayed with 2,4-D.

The explanation of this peculiar behaviour sometimes appears to lie in the changes which the chemical brings about in the metabolism of the plant itself. There is temporarily a marked increase in sugar content, making the plant more attractive to many animals.

Another curious effect of 2,4-D has important effects for livestock, wildlife, and apparently for men as well. Experiments carried out about a decade ago showed that after treatment with this chemical there is a sharp increase in the nitrate content of corn and sugar beets. The same effect was suspected in sorghum, sunflower, spiderwort, lambs quarters, pigweed, and smartweed. Some of these are normally ignored by cattle, but are eaten with relish after treatment with 2,4-D. A number of deaths among cattle have been traced to sprayed weeds, according to some agricultural specialists. The danger lies in the increase in nitrates, for the peculiar physiology of the ruminant at once poses a critical problem. Most such animals have a digestive system of extraordinary complexity, including a stomach divided into four chambers. The digestion of cellulose is accomplished through the action of micro-organisms (rumen bacteria) in one of the chambers. When the animal feeds on vegetation containing an abnormally high level of nitrates, the micro-organisms in the rumen act on the nitrates to change them into highly toxic nitrites. Thereafter a fatal chain of events ensues: the nitrites act on the blood pigment to form a chocolate-brown substance in which the oxygen is so firmly held that it cannot take part in respiration, hence oxygen is not transferred from the lungs to the tissues. Death occurs within a few hours from anoxia, or lack of oxygen. The various reports of livestock losses after grazing on certain weeds treated with 2,4-D therefore have a logical explanation. The same danger exists for wild animals belonging to the group of ruminants, such as deer, antelope, sheep, and goats.

Although various factors (such as exceptionally dry weather) can cause an increase in nitrate content, the effect of the soaring sales and applications of 2,4-D cannot be ignored. The situation was considered important enough by the University of Wisconsin Agricultural Experiment Station to justify a warning in 1957 that 'plants killed by 2,4-D may contain large amounts of nitrate'. The hazard extends to human beings as well as animals and may help to explain the recent mysterious increase in 'silo deaths'. When corn, oats, or sorghum containing large amounts of nitrates are ensiled they release poisonous nitrogen oxide gases, creating a deadly hazard to anyone entering the silo. Only a few breaths of one of these gases can cause a diffuse chemical pneumonia. In a series of such cases studied by the University of Minnesota Medical School all but one terminated fatally.

'Once again we are walking in nature like an elephant in the china cabinet.' So C. J. Briejèr, a Dutch scientist of rare understanding, sums up our use of weed killers. 'In my opinion too much is taken for granted. We do not know whether all weeds in crops are harmful or whether some of them are useful,' says Dr Briejèr.

Seldom is the question asked, What is the relation between the weed and the soil? Perhaps, even from our narrow standpoint of direct self-interest, the relation is a useful one. As we have seen, soil and the living things in and upon it exist in a relation of interdependence and mutual benefit. Presumably the weed is taking something from the soil; perhaps it is also contributing something to it. A practical example was provided recently by the parks in a city in Holland. The roses were doing badly. Soil samples showed heavy infestations by tiny nematode worms. Scientists of the Dutch Plant Protection Service did not recommend chemical sprays or soil treatments; instead, they suggested that marigolds be planted among the roses. This plant, which the purist would doubtless consider a weed in any rosebed, releases an excretion from its roots that kills the soil nematodes. The advice was taken; some beds were planted with marigolds, some left without as controls. The results were striking. With the aid of the marigolds the roses flourished; in the control beds they were sickly and drooping. Marigolds are now used in many places for combating nematodes.

In the same way, and perhaps quite unknown to us, other plants that we ruthlessly eradicate may be performing a function that is necessary to the health of the soil. One very useful function of natural plant communities – now pretty generally stigmatized as 'weeds' – is to serve as an indicator of the condition of the soil. This useful function is of course lost where chemical weed killers have been used.

Those who find an answer to all problems in spraying also overlook a matter of great scientific importance – the need to preserve some natural plant communities. We need these as a standard against which we can measure the changes our own activities bring about. We need them as wild habitats in which original populations of insects and other organisms can be maintained, for, as will be explained in Chapter 16, the development of resistance to insecticides is changing the genetic factors of insects and perhaps other organisms. One scientist has even suggested that some sort of 'zoo' should be established to preserve insects, mites, and the like, before their genetic composition is further changed.

Some experts warn of subtle but far-reaching vegetational shifts as a result of the growing use of herbicides. The chemical 2,4-D, by killing out the broad-leaved plants, allows the grasses to thrive in the reduced competition – now some of the grasses themselves have become 'weeds', presenting a new problem in control and giving the cycle another turn. This strange situation is acknowledged in a recent issue of a journal devoted to crop problems:

> With the widespread use of 2,4-D to control broad-leaved weeds, grass weeds in particular have increasingly become a threat to corn and soybean yields.

Ragweed, the bane of hay-fever sufferers, offers an interesting example of the way efforts to control nature sometimes boomerang. Many thousands of gallons of chemicals have been discharged along roadsides in the name of ragweed control. But the unfortunate truth is that blanket spraying is resulting in more ragweed, not less. Ragweed is an annual; its seedlings require open soil to become established each year. Our best protection against this plant is therefore the maintenance of dense shrubs, ferns, and other perennial vegetation. Spraying frequently destroys this protective vegetation and creates open, barren areas which the ragweed hastens to fill. It is probable, moreover, that the pollen content of the atmosphere is not related to roadside ragweed, but to the ragweed of city lots and fallow fields.

The booming sales of chemical crabgrass killers are another example of how readily unsound methods catch on. There is a cheaper and better way to remove crabgrass than to attempt year after year to kill it out with chemicals. This is to give it competition of a kind it cannot survive, the competition of other grass. Crabgrass exists only in an unhealthy lawn. It is a symptom, not a disease in itself. By providing a fertile soil and giving the desired grasses a good start, it is possible to create an environment in which crabgrass cannot grow, for it requires open space in which it can start from seed year after year.

Instead of treating the basic condition, suburbanites – advised by nurserymen who in turn have been advised by the chemical manufacturers – continue to apply truly astonishing amounts of crabgrass killers to their lawns each year. Marketed under trade names which give no hint of their nature, many of these preparations contain such poisons as mercury, arsenic, and chlordane. Application at the recommended rate leaves tremendous amounts of these chemicals on the lawn. Users of one product, for example, apply sixty pounds of technical chlordane to the acre if they follow directions. If they use another of the many available products, they are applying 175 pounds of metallic arsenic to the acre. The toll of dead birds, as we shall see in Chapter 8, is distressing. How lethal these lawns may be for human beings is unknown.

The success of selective spraying for roadside and right-of-way vegetation, where it has been practised, offers hope that equally sound ecological methods may be developed for other vegetation programmes for farms, forests, and ranges – methods aimed not at destroying a particular species but at managing vegetation as a living community.

Other solid achievements show what can be done. Biological control has achieved some of its most spectacular successes in the area of curbing unwanted vegetation. Nature herself has met many of the problems that now beset us, and she has usually solved them in her own successful way. Where man has been intelligent enough to observe and to emulate nature he, too, is often rewarded with success.

An outstanding example in the field of controlling unwanted plants is the handling of the Klamath-weed problem in California. Although the Klamath weed, or goatweed, is a native of Europe (where it is called St John's wort), it accompanied man in his westward migrations, first appearing in the United States in 1793 near Lancaster, Pennsylvania. By 1900 it had reached California in the vicinity of the Klamath River, hence the name locally given to it. By 1929 it had occupied about 100,000 acres of rangeland, and by 1952 it had invaded some two and a half million acres.

Klamath weed, quite unlike such native plants as sagebrush, has no place in the ecology of the region, and no animals or other plants require its presence. On the contrary, wherever it appeared, livestock became 'scabby, sore-mouthed, and un-thrifty' from feeding on this toxic plant. Land values declined accordingly, for the Klamath weed was considered to hold the first mortgage.

In Europe the Klamath weed, or St John's wort, has never become a problem because along with the plant there have developed various species of insects; these feed on it so extensively that its abundance is severely limited. In particular, two species of beetles in southern France, pea-sized and of metallic colour, have their whole beings so adapted to the presence of the weed that they feed and reproduce only upon it.

It was an event of historic importance when the first shipments of these beetles were brought to the United States in 1944, for this was the first attempt in North America to control a plant with a plant-eating insect. By 1948 both species had become so well established that no further importations were needed. Their spread was accomplished by collecting beetles from the original colonies and redistributing them at the rate of millions a year. Within small areas the beetles accomplish their own dispersion, moving on as soon as the Klamath weed dies out and locating new stands with great precision. And as the beetles thin out the weed, desirable range plants that have been crowded out are able to return.

A ten-year survey completed in 1959 showed that control of the Klamath weed had been 'more effective than hoped for even by enthusiasts', with the weed reduced to a mere 1 per cent of its former abundance. This token infestation is harmless and is actually needed in order to maintain a population of beetles as protection against a future increase in the weed.

Another extraordinarily successful and economical example of weed control may be found in Australia. With the colonists' usual taste for carrying plants or animals into a new country, a Captain Arthur Phillip had brought various species of cactus into Australia about 1787, intending to use them in culturing cochineal insects for dye. Some of the cacti or prickly pears escaped from his gardens and by 1925 about twenty species could be found growing wild. Having no natural controls in this new territory, they spread prodigiously, eventually occupying about sixty million acres. At least half of this land was so densely covered as to be useless.

In 1920 Australian entomologists were sent to North and South America to study insect enemies of the prickly pears in their native habitat. After trials of several species, three billion eggs of an Argentine moth were released in Australia in 1930. Seven years later the last dense growth of the prickly pear had been destroyed and the once uninhabitable areas reopened to settlement and grazing. The whole operation had cost less than a penny per acre. In contrast, the unsatisfactory attempts at chemical control in earlier years had cost about £10 per acre.

Both of these examples suggest that extremely effective control of many kinds of unwanted vegetation might be achieved by paying more attention to the role of plant-eating insects. The science of range management has largely ignored this possibility, although these insects are perhaps the most selective of all grazers and their highly restricted diets could easily be turned to man's advantage.

Chapter 7

Needless Havoc

As man proceeds towards his announced goal of the conquest of nature, he has written a depressing record of destruction, directed not only against the earth he inhabits but against the life that shares it with him. The history of the recent centuries has its black passages – the slaughter of the buffalo on the western plains, the massacre of the shore-birds by the market gunners, the near-extermination of the egrets for their plumage. Now, to these and others like them, we are adding a new chapter and a new kind of havoc – the direct killing of birds, mammals, fishes, and indeed practically every form of wildlife by chemical insecticides indiscriminately sprayed on the land.

Under the philosophy that now seems to guide our destinies, nothing must get in the way of the man with the spray gun. The incidental victims of his crusade against insects count as nothing; if robins, pheasants, raccoons, cats, or even livestock happen to inhabit the same bit of earth as the target insects and to be hit by the rain of insect-killing poisons no one must protest.

The citizen who wishes to make a fair judgement of the question of wildlife loss is today confronted with a dilemma. On the one hand conservationists and many wildlife biologists assert that the losses have been severe and in some cases even catastrophic. On the other hand the control agencies tend to deny flatly and categorically that such losses have occurred, or that they are of any importance if they have. Which view are we to accept?

The credibility of the witness is of first importance. The professional wildlife biologist on the scene is certainly best qualified to discover and interpret wildlife loss. The entomologist, whose speciality is insects, is not so qualified by training, and is not psychologically disposed to look for undesirable side effects of his control programme. Yet it is the control men in state and federal governments – and of course the chemical manufacturers – who steadfastly deny the facts reported by the biologists and declare they see little evidence of harm to wildlife. Like the priest and the Levite in the biblical story, they choose to pass by on the other side and to see nothing. Even if we charitably explain their denials as due to shortsightedness of the specialist and the man with an interest this does not mean we must accept them as qualified witnesses.

The best way to form our own judgement is to look at some of the major control programmes and learn, from observers familiar with the ways of wildlife, and unbiased in favour of chemicals, just what has happened in the wake of a rain of poison falling from the skies into the world of wildlife.

To the bird watcher, the suburbanite who derives joy from birds in his garden, the hunter, the fisherman or the explorer of wild regions, anything that destroys the wildlife of an area for even a single year has deprived him of pleasure to which he has a legitimate right. This is a valid point of view. Even if, as has sometimes happened, some of the birds and mammals and fishes are able to re-establish themselves after a single spraying, a great and real harm has been done.

But such re-establishment is unlikely to happen. Spraying tends to be repetitive, and a single exposure from which the wildlife populations might have a chance to recover is a rarity. What usually results is a poisoned environment, a lethal trap in which not only the resident population succumb but those who come in as migrants as well. The larger the area sprayed the more serious the harm, because no oases of safety remain. Now, in a decade marked by insect-control programmes in which many thousands or even millions of acres are sprayed as a unit, a decade in which private and community spraying has also surged steadily upwards, a record of destruction and death of American wildlife has accumulated. Let us look at some of these programmes and see what has happened.

During the autumn of 1959 some 27,000 acres in south-eastern Michigan, including numerous suburbs of Detroit, were heavily dusted from the air with pellets of aldrin, one of the most dangerous of all the chlorinated hydrocarbons. The programme was conducted by the Michigan Department of Agriculture with the cooperation of the United States Department of Agriculture; its announced purpose was control of the Japanese beetle.

Little need was shown for this drastic and dangerous action. On the contrary, Walter P. Nickell, one of the best-known and best-informed naturalists in the state, who spends much of his time in the field with long periods in southern Michigan every summer, declared:

For more than thirty years, to my direct knowledge, the Japanese beetle has been present in the city of Detroit in small numbers. The numbers have not shown any appreciable increase in all this lapse of years. I have yet to see a single Japanese beetle [in 1959] other than the few caught in Government catch traps in Detroit.... Everything is being kept so secret that I have not yet been able to obtain any information whatsoever to the effect that they have increased in numbers.

An official release by the state agency merely declared that the beetle had 'put in its appearance' in the areas designated for the aerial attack upon it. Despite the lack of justification the programme was launched, with the state providing the manpower and supervising the operation, the federal government providing equipment and additional men, and the communities paying for the insecticide.

The Japanese beetle, an insect accidentally imported into the United States, was discovered in New Jersey in 1916, when a few shiny beetles of a metallic green colour were seen in a nursery near Riverton. The beetles, at first unrecognized, were finally identified as a common inhabitant of the main islands of Japan. Apparently they had entered the United States on nursery stock imported before restrictions were established in 1912.

From its original point of entrance the Japanese beetle has spread rather widely throughout many of the states east of the Mississippi, where conditions of temperature and rainfall are suitable for it. Each year some outward movement beyond the existing boundaries of its distribution usually takes place. In the eastern areas where the beetles have been longest established, attempts have been made to set up natural controls. Where this has been done, the beetle populations have been kept at relatively low levels, as many records attest.

Despite the record of reasonable control in eastern areas, the mid-western states now on the fringe of the beetle's range have launched an attack worthy of the most deadly enemy instead of only a moderately destructive insect, employing the most dangerous chemicals distributed in a manner that exposes large numbers of people, their domestic animals, and all wildlife to the poison intended for the beetle. As a result these Japanese beetle programmes have caused shocking destruction of animal life and have exposed human beings to undeniable hazard. Sections of Michigan, Kentucky, Iowa, Indiana, Illinois, and Missouri are all experiencing a rain of chemicals in the name of beetle control.

The Michigan spraying was one of the first large-scale attacks on the Japanese beetle from the air. The choice of aldrin, one of the deadliest of all chemicals, was not determined by any peculiar suitability for Japanese beetle control, but simply by the wish to save money – aldrin was the cheapest of the compounds available. While the state in its official release to the press acknowledged that aldrin is a 'poison', it implied that no harm could come to human beings in the heavily populated areas to which the chemical was applied. (The official answer to the query 'What precautions should I take?' was 'For you, none.') An official of the Federal Aviation Agency was later quoted in the local press to the effect that 'this is a safe operation' and a representative of the Detroit Department of Parks and Recreation added his assurance that 'the dust is harmless to humans and will not hurt plants or pets'. One must assume that none of these officials had consulted the published and readily available reports of the United States Public Health Service, the Fish and Wildlife Service, and other evidence of the extremely poisonous nature of aldrin.

Acting under the Michigan pest control law which allows the state to spray indiscriminately without notifying or gaining permission of individual landowners, the low-flying planes began to fly over the Detroit area. The city authorities and the Federal Aviation Agency were immediately besieged by calls from worried citizens. After receiving nearly 800 calls in a single hour, the police begged radio and television stations and newspapers to 'tell the watchers what they were seeing and advise them it was safe', according to the Detroit News. The Federal Aviation Agency's safety officer assured the public that 'the planes are carefully supervised' and 'are authorized to fly low'. In a somewhat mistaken attempt to allay fears, he added that the planes had emergency valves that would allow them to dump their entire load instantaneously. This, fortunately, was not done, but as the planes went about their work the pellets of insecticide fell on beetles and humans alike, showers of 'harmless' poison descending on people shopping or going to work and on children out from school for the lunch hour. Housewives swept the granules from porches and pavements, where they are said to have 'looked like snow'. As pointed out later by the Michigan Audubon Society,

In the spaces between shingles on roofs, in eaves-troughs, in the cracks in bark and twigs, the little white pellets of aldrin-and-clay, no bigger than a pin head, were lodged by the millions.... When the snow and rain came, every puddle became a possible death potion.

Within a few days after the dusting operation, the Detroit Audubon Society began receiving calls about the birds. According to the Society's secretary, Mrs Ann Boyes,

The first indication that the people were concerned about the spray was a call I received on Sunday morning from a woman who reported that coming home from church she saw an alarming number of dead and dying birds. The spraying there had been done on Thursday. She said there were no birds at all flying in the area, that she had found at least a dozen [dead] in her backyard and that the neighbours had found dead squirrels.

All other calls received by Mrs Boyes that day reported 'a great many dead birds and no live ones.... People who had maintained bird feeders said there were no birds at all at their feeders.' Birds picked up in a dying condition showed the typical symptoms of insecticide poisoning – tremoring, loss of ability to fly, paralysis, convulsions.

Nor were birds the only forms of life immediately affected. A local veterinarian reported that his office was full of clients with dogs and cats that had suddenly sickened. Cats, who so meticulously groom their coats and lick their paws, seemed to be most affected. Their illness took the form of severe diarrhoea, vomiting, and convulsions. The only advice the veterinarian could give his clients was not to let the animals out unnecessarily, or to wash the paws promptly if they did so. (But the chlorinated hydrocarbons cannot be washed even from fruits or vegetables, so little protection could be expected from this measure.)

Despite the insistence of the City-County Health Commissioner that the birds must have been killed by 'some other kind of spraying' and that the outbreak of throat and chest irritations that followed the exposure to aldrin must have been due to 'something else', the local Health Department received a constant stream of complaints. A prominent Detroit internist was called upon to treat four of his patients within an hour after they had been exposed while watching the planes at work. All had similar symptoms: nausea, vomiting, chills, fever, extreme fatigue, and coughing.

The Detroit experience has been repeated in many other communities as pressure has mounted to combat the Japanese beetle with chemicals. At Blue Island, Illinois, hundreds of dead and dying birds were picked up. Data collected by bird-banders here suggest that 80 per cent of the songbirds were sacrificed. In Joliet, Illinois, some 3,000 acres were treated with heptachlor in 1959. According to reports from a local sportsmen's club, the bird population within the treated area was 'virtually wiped out'. Dead rabbits, muskrats, opossums, and fish were also found in numbers, and one of the local schools made the collection of insecticide-poisoned birds a science project.

Perhaps no community has suffered more for the sake of a beetleless world than Sheldon, in eastern Illinois, and adjacent areas in Iroquois County. In 1954 the United States Department of Agriculture and the Illinois Agriculture Department began a programme to eradicate the Japanese beetle along the line of its advance into Illinois, holding out the hope, and indeed the assurance, that intensive spraying would destroy the populations of the invading insect. The first 'eradication' took place that year, when dieldrin was applied to 1,400 acres by air. Another 2,600 acres were treated similarly in 1955, and the task was presumably considered complete. But more and more chemical treatments were called for, and by the end of 1961 some 131,000 acres had been covered. Even in the first years of the programme it was apparent that heavy losses were occurring among wildlife and domestic animals. The chemical treatments were continued, nevertheless, without consultation with either the United States Fish and Wildlife Service or the Illinois Game Management Division. (In the spring of 1960, however, officials of the federal Department of Agriculture appeared before a congressional committee in opposition to a bill that would require just such prior consultation. They declared blandly that the bill was unnecessary because cooperation and consultation were 'usual'. These officials were quite unable to recall situations where cooperation had not taken place 'at the Washington level'. In the same hearings they stated clearly their unwillingness to consult with state fish and game departments.)

Although funds for chemical control came in never-ending streams, the biologists of the Illinois Natural History Survey who attempted to measure the damage to wildlife had to operate on a financial shoestring. A mere $1,100 was available for the employment of a field assistant in 1954 and no special funds were provided in 1955. Despite these crippling difficulties, the biologists assembled facts that collectively paint a picture of almost unparalleled wildlife destruction – destruction that became obvious as soon as the programme got under way.

Conditions were made to order for poisoning insect-eating birds, both in the poisons used and in the events set in motion by their application. In the early programmes at Sheldon, dieldrin was applied at the rate of 3 pounds to the acre. To understand its effect on birds one need only remember that in laboratory experiments on quail dieldrin has proved to be about fifty times as poisonous as DDT. The poison spread over the landscape at Sheldon was therefore roughly equivalent to 150 pounds of DDT per acre! And this was a minimum, because there seems to have been some overlapping of treatments along field borders and in corners.

As the chemical penetrated the soil the poisoned beetle grubs crawled out on the surface of the ground, where they remained for some time before they died, attractive to insect-eating birds. Dead and dying insects of various species were conspicuous for about two weeks after the treatment. The effect on the bird populations could easily have been foretold. Brown thrashers, starlings, meadow larks, grackles, and pheasants were virtually wiped out. Robins were 'almost annihilated', according to the biologists' report. Dead earthworms had been seen in numbers after a gentle rain; probably the robins had fed on the poisoned worms. For other birds, too, the once beneficial rain had been changed, through the evil power of the poison introduced into their world, into an agent of destruction. Birds seen drinking and bathing in puddles left by rain a few days after the spraying were inevitably doomed.

The birds that survived may have been rendered sterile. Although a few nests were found in the treated area, a few with eggs, none contained young birds.

Among the mammals ground squirrels were virtually annihilated; their bodies were found in attitudes characteristic of violent death by poisoning. Dead muskrats were found in the treated areas, dead rabbits in the fields. The fox squirrel had been a relatively common animal in the town; after the spraying it was gone.

It was a rare farm in the Sheldon area that was blessed by the presence of a cat after the war on beetles was begun. Ninety per cent of all the farm cats fell victims to the dieldrin during the first season of spraying. This might have been predicted because of the black record of these poisons in other places. Cats are extremely sensitive to all insecticides and especially so, it seems, to dieldrin. In western Java in the course of the anti-malarial programme carried out by the World Health Organiza-tion, many cats are reported to have died. In central Java so many were killed that the price of a cat more than doubled. Similarly, the World Health Organization, spraying in Vene-zuela, is reported to have reduced cats to the status of a rare animal.

In Sheldon it was not only the wild creatures and the domestic companions that were sacrificed in the campaign against an insect. Observations on several flocks of sheep and a herd of beef cattle are indicative of the poisoning and death that threa-tened livestock as well. The Natural History Survey report describes one of these episodes as follows:

The sheep ... were driven into a small, untreated blue-grass pasture across a gravel road from a field which had been treated with dieldrin spray on May 6. Evidently some spray had drifted across the road into the pasture, for the sheep began to show symptoms of intoxication almost at once.... They lost interest in food and displayed extreme restlessness, following the pasture fence around and around apparently searching for a way out.... [They] refused to be driven, bleated almost continuously, and stood with their heads lowered; they were finally carried from the pasture.... They displayed great desire for water. Two of the sheep were found dead in the stream passing through the pasture, and the remaining sheep were repeatedly driven out of the stream, several having to be dragged forcibly from the water. Three of the sheep eventually died; those remaining recovered to all outward appearances.

This, then, was the picture at the end of 1955. Although the chemical war went on in succeeding years, the trickle of research funds dried up completely. Requests for money for wildlife-insecticide research were included in annual budgets submitted to the Illinois legislature by the Natural History Survey, but were invariably among the first items to be eliminated. It was not until 1960 that money was somehow found to pay the expenses of one field assistant – to do work that could easily have occupied the time of four men.

The desolate picture of wildlife loss had changed little when the biologists resumed the studies broken off in 1955. In the meantime, the chemical had been changed to the even more toxic aldrin, *100 to 300 times* as toxic as DDT in tests on quail. By 1960, every species of wild mammal known to inhabit the area had suffered losses. It was even worse with the birds. In the small town of Donovan the robins had been wiped out, as had the grackles, starlings, and brown thrashers. These and many other birds were sharply reduced elsewhere. Pheasant hunters felt the effects of the beetle campaign sharply. The number of broods produced on treated lands fell off by some 50 per cent, and the number of young in a brood declined. Pheasant hunting, which had been good in these areas in former years, was virtually abandoned as unrewarding.

In spite of the enormous havoc that had been wrought in the name of eradicating the Japanese beetle, the treatment of more than 100,000 acres in Iroquois County over an eight-year period seems to have resulted in only temporary suppression of the insect, which continues its westward movement. The full extent

of the toll that has been taken by this largely ineffective programme may never be known, for the results measured by the
Illinois biologists are a minimum figure. If the research programme had been adequately financed to permit full coverage
the destruction revealed would have been even more appalling.
But in the eight years of the programme, only about $6,000 was
provided for biological field studies. Meanwhile the federal
government had spent about $375,000 for control work and
additional thousands had been provided by the state. The amount
spent for research was therefore a small fraction of 1 per cent of
the outlay for the chemical programme.

These mid-western programmes have been conducted in a
spirit of crisis, as though the advance of the beetle presented an
extreme peril justifying any means to combat it. This, of course,
is a distortion of the facts, and if the communities that have
endured these chemical drenchings had been familiar with the
earlier history of the Japanese beetle in the United States they
would surely have been less acquiescent.

The eastern states, which had the good fortune to sustain their
beetle invasion in the days before the synthetic insecticides had
been invented, have not only survived the invasion but have
brought the insect under control by means that represented no
threat whatever to other forms of life. There has been nothing
comparable to the Detroit or Sheldon sprayings in the East.
The effective methods there involved the bringing into play of
natural forces of control which have the multiple advantages of
permanence and environmental safety.

During the first dozen years after its entry into the United
States, the beetle increased rapidly, free of the restraints that in
its native land hold it in check. But by 1945 it had become a pest
of only minor importance throughout much of the territory over
which it had spread. Its decline was largely a consequence of the
importation of parasitic insects from the Far East and of the
establishment of disease organisms fatal to it.

Between 1920 and 1933, as a result of diligent searching
throughout the native range of the beetle, some thirty-four
species of predatory or parasitic insects had been imported from
the Orient in an effort to establish natural control. Of these, five
became well established in the eastern United States. The most

effective and widely distributed is a parasitic wasp from Korea and China, *Tiphia vernalis*. The female *Tiphia*, finding a beetle grub in the soil, injects a paralysing fluid and attaches a single egg to the under-surface of the grub. The young wasp, hatching as a larva, feeds on the paralysed grub and destroys it. In some twenty-five years, colonies of *Tiphia* were introduced into four-teen eastern states in a cooperative programme of state and federal agencies. The wasp became widely established in this area and is generally credited by entomologists with an important role in bringing the beetle under control.

An even more important role has been played by a bacterial disease that affects beetles of the family to which the Japanese beetle belongs – the scarabaeids. It is a highly specific organism, attacking no other type of insects, harmless to earthworms, warm-blooded animals, and plants. The spores of the disease occur in soil. When ingested by a foraging beetle grub they multiply prodigiously in its blood, causing it to turn an abnormally white colour, hence the popular name, 'milky disease'.

Milky disease was discovered in New Jersey in 1933. By 1938 it was rather widely prevalent in the older areas of Japanese beetle infestation. In 1939 a control programme was launched, directed at speeding up the spread of the disease. No method had been developed for growing the disease organism in an artificial medium, but a satisfactory substitute was evolved; infected grubs are ground up, dried, and combined with chalk. In the standard mixture a gram of dust contains 100 million spores. Between 1939 and 1953 some 94,000 acres in fourteen eastern states were treated in a cooperative federal-state programme; other areas on federal lands were treated; and an unknown but extensive area was treated by private organizations or individuals. By 1945, milky spore disease was raging among the beetle populations of Connecticut, New York, New Jersey, Delaware, and Maryland. In some test areas infection of grubs had reached as high as 94 per cent. The distribution programme was dis-continued as a governmental enterprise in 1953 and production was taken over by a private laboratory, which continues to supply individuals, garden clubs, citizens' associations, and all others interested in beetle control.

The eastern areas where this programme was carried out now

enjoy a high degree of natural protection from the beetle. The organism remains viable in the soil for years and therefore becomes to all intents and purposes permanently established, increasing in effectiveness, and being continuously spread by natural agencies.

Why, then, with this impressive record in the East, were the same procedures not tried in Illinois and the other mid-western states where the chemical battle of the beetles is now being waged with such fury?

We are told that inoculation with milky spore disease is 'too expensive' – although no one found it so in the fourteen eastern states in the 1940s. And by what sort of accounting was the 'too expensive' judgement reached? Certainly not by any that assessed the true costs of the total destruction wrought by such programmes as the Sheldon spraying. This judgement also ignores the fact that inoculation with the spores need be done only once; the first cost is the only cost.

We are told also that milky spore disease cannot be used on the periphery of the beetle's range because it can be established only where a large grub population is *already* present in the soil. Like many other statements in support of spraying, this one needs to be questioned. The bacterium that causes milky spore disease has been found to infect at least forty other species of beetles which collectively have quite a wide distribution and would in all probability serve to establish the disease even where the Japanese beetle population is very small or non-existent. Furthermore, because of the long viability of the spores in soil they can be introduced even in the complete absence of grubs, as on the fringe of the present beetle infestation, there to await the advancing population.

Those who want immediate results, at whatever cost, will doubtless continue to use chemicals against the beetle. So will those who favour the modern trend to built-in obsolescence, for chemical control is self-perpetuating, needing frequent and costly repetition.

On the other hand, those who are willing to wait an extra season or two for full results will turn to milky disease; they will be rewarded with lasting control that becomes more, rather than less, effective with the passage of time.

An extensive programme of research is under way in the United States Department of Agriculture laboratory at Peoria, Illinois, to find a way to culture the organism of milky disease on an artificial medium. This will greatly reduce its cost and should encourage its more extensive use. After years of work, some success has now been reported. When this 'break-through' is thoroughly established perhaps some sanity and perspective will be restored to our dealings with the Japanese beetle, which at the peak of its depredations never justified the nightmare excesses of some of these mid-western programmes.

Incidents like the eastern Illinois spraying raise a question that is not only scientific but moral. The question is whether any civilization can wage relentless war on life without destroying itself, and without losing the right to be called civilized.

These insecticides are not selective poisons; they do not single out the one species of which we desire to be rid. Each of them is used for the simple reason that it is a deadly poison. It therefore poisons all life with which it comes in contact: the cat beloved of some family, the farmer's cattle, the rabbit in the field, and the horned lark out of the sky. These creatures are innocent of any harm to man. Indeed, by their very existence they and their fellows make his life more pleasant. Yet he rewards them with a death that is not only sudden but horrible. Scientific observers at Sheldon described the symptoms of a meadow lark found near death:

Although it lacked muscular co-ordination and could not fly or stand, it continued to beat its wings and clutch with its toes while lying on its side. Its beak was held open and breathing was laboured.

Even more pitiful was the mute testimony of the dead ground squirrels, which

exhibited a characteristic attitude in death. The back was bowed, and the forelegs with the toes of the feet tightly clenched were drawn close to the thorax.... The head and neck were outstretched and the mouth often contained dirt, suggesting that the dying animal had been biting at the ground.

By acquiescing in an act that can cause such suffering to a living creature, who among us is not diminished as a human being?

Chapter 8

And No Birds Sing

Over increasingly large areas of the United States, spring now comes unheralded by the return of the birds, and the early mornings are strangely silent where once they were filled with the beauty of bird song. This sudden silencing of the song of birds, this obliteration of the colour and beauty and interest they lend to our world have come about swiftly, insidiously, and unnoticed by those whose communities are as yet unaffected.

From the town of Hinsdale, Illinois, a housewife wrote in despair to one of the world's leading ornithologists, Robert Cushman Murphy, Curator Emeritus of Birds at the American Museum of Natural History.

Here in our village the elm trees have been sprayed for several years [she wrote in 1958]. When we moved here six years ago, there was a wealth of bird life; I put up a feeder and had a steady stream of cardinals, chickadees, downies and nuthatches all winter, and the cardinals and chickadees brought their young ones in the summer.

After several years of DDT spray, the town is almost devoid of robins and starlings; chickadees have not been on my shelf for two years, and this year the cardinals are gone too; the nesting population in the neighbourhood seems to consist of one dove pair and perhaps one catbird family.

It is hard to explain to the children that the birds have been killed off, when they have learned in school that a Federal law protects the birds from killing or capture. 'Will they ever come back?' they ask, and I do not have the answer. The elms are still dying, and so are the birds. *Is* anything being done? *Can* anything be done? Can *I* do anything?

A year after the federal government had launched a massive spraying programme against the fire ant, an Alabama woman wrote:

Our place has been a veritable bird sanctuary for over half a century.

Last July we all remarked, 'There are more birds than ever.' Then, suddenly, in the second week of August, they all disappeared. I was accustomed to rising early to care for my favourite mare that had a young filly. There was not a sound of the song of a bird. It was eerie, terrifying. What was man doing to our perfect and beautiful world? Finally, five months later a blue jay appeared and a wren.

The autumn months to which she referred brought other sombre reports from the deep South, where in Mississippi, Louisiana, and Alabama the *Field Notes* published quarterly by the National Audubon Society and the United States Fish and Wildlife Service noted the striking phenomenon of 'blank spots weirdly empty of virtually *all* bird life'. The *Field Notes* are a compilation of the reports of seasoned observers who have spent many years afield in their particular areas and have unparalleled knowledge of the normal bird life of the region. One such observer reported that in driving about southern Mississippi that autumn she saw 'no land birds at all for long distances'. Another in Baton Rouge reported that the contents of her feeders had lain untouched 'for weeks on end', while fruiting shrubs in her yard, that ordinarily would be stripped clean by that time, still were laden with berries. Still another reported that his picture window, 'which often used to frame a scene splashed with the red of forty or fifty cardinals and crowded with other species, seldom permitted a view of as many as a bird or two at a time'. Professor Maurice Brooks of the University of West Virginia, an authority on the birds of the Appalachian region, reported that the West Virginia bird population had undergone 'an incredible reduction'.

One story might serve as the tragic symbol of the fate of the birds – a fate that has already overtaken some species, and that threatens all. It is the story of the robin, the bird known to everyone. To millions of Americans, the season's first robin means that the grip of winter is broken. Its coming is an event reported in newspapers and told eagerly at the breakfast table. And as the number of migrants grows and the first mists of green appear in the woodlands, thousands of people listen for the first dawn chorus of the robins throbbing in the early morning light. But now all is changed, and not even the return of the birds may be taken for granted.

The survival of the robin, and indeed of many other species as well, seems fatefully linked with the American elm, a tree that is part of the history of thousands of towns from the Atlantic to the Rockies, gracing their streets and their village squares and college campuses with majestic archways of green. Now the elms are stricken with a disease that afflicts them throughout their range, a disease so serious that many experts believe all efforts to save the elms will in the end be futile. It would be tragic to lose the elms, but it would be doubly tragic if, in vain efforts to save them, we plunge vast segments of our bird populations into the night of extinction. Yet this is precisely what is threatened.

The so-called Dutch elm disease entered the United States from Europe about 1930 in elm burl logs imported for the veneer industry. It is a fungus disease; the organism invades the water-conducting vessels of the tree, spreads by spores carried in the flow of sap, and by its poisonous secretions as well as by mechanical clogging causes the branches to wilt and the tree to die. The disease is spread from diseased to healthy trees by elm bark beetles. The galleries which the insects have tunnelled out under the bark of dead trees become contaminated with spores of the invading fungus, and the spores adhere to the insect body and are carried wherever the beetle flies. Efforts to control the fungus disease of the elms have been directed largely towards control of the carrier insect. In community after community, especially throughout the strongholds of the American elm, the Midwest and New England, intensive spraying has become a routine procedure.

What this spraying could mean to bird life, and especially to the robin, was first made clear by the work of two ornithologists at Michigan State University, Professor George Wallace and one of his graduate students, John Mehner. When Mr Mehner began work for the doctorate in 1954, he chose a research project that had to do with robin populations. This was quite by chance, for at that time no one suspected that the robins were in danger. But even as he undertook the work, events occurred that were to change its character and indeed to deprive him of his material.

Spraying for Dutch elm disease began in a small way on the university campus in 1954. The following year the city of East

Lansing (where the university is located) joined in, spraying on the campus was expanded, and, with local programmes for gypsy moth and mosquito control also under way, the rain of chemicals increased to a downpour.

During 1954, the year of the first light spraying, all seemed well. The following spring the migrating robins began to return to the campus as usual. Like the bluebells in Tomlinson's haunting essay 'The Lost Wood', they were 'expecting no evil' as they reoccupied their familiar territories. But soon it became evident that something was wrong. Dead and dying robins began to appear on the campus. Few birds were seen in their normal foraging activities or assembling in their usual roosts. Few nests were built; few young appeared. The pattern was repeated with monotonous regularity in succeeding springs. The sprayed area had become a lethal trap in which each wave of migrating robins would be eliminated in about a week. Then new arrivals would come in, only to add to the numbers of doomed birds seen on the campus in the agonized tremors that precede death.

'The campus is serving as a graveyard for most of the robins that attempt to take up residence in the spring,' said Dr Wallace. But why? At first he suspected some disease of the nervous system, but soon it became evident that

in spite of the assurances of the insecticide people that their sprays were 'harmless to birds' the robins were really dying of insecticidal poisoning; they exhibited the well-known symptoms of loss of balance, followed by tremors, convulsions, and death.

Several facts suggested that the robins were being poisoned, not so much by direct contact with the insecticides as indirectly, by eating earthworms. Campus earthworms had been fed inadvertently to crayfish in a research project and all the crayfish had promptly died. A snake kept in a laboratory cage had gone into violent tremors after being fed such worms. And earthworms are the principal food of robins in the spring.

A key piece in the jigsaw puzzle of the doomed robins was soon to be supplied by Dr Roy Barker of the Illinois Natural History Survey at Urbana. Dr Barker's work, published in 1958, traced the intricate cycle of events by which the robins' fate is linked to the elm trees by way of the earthworms. The trees are sprayed in the spring (usually at the rate of 2 to 6 pounds of DDT per 50-foot tree, which may be the equivalent of as much as 23 *pounds per acre* where elms are numerous) and often again in July, at about half this concentration. Powerful sprayers direct a stream of poison to all parts of the tallest trees, killing directly not only the target organism, the bark beetle, but other insects, including pollinating species and predatory spiders and beetles. The poison forms a tenacious film over the leaves and bark. Rains do not wash it away. In the autumn the leaves fall to the ground, accumulate in sodden layers, and begin the slow process of becoming one with the soil. In this they are aided by the toil of the earthworms, who feed in the leaf litter, for elm leaves are among their favourite foods. In feeding on the leaves the worms always swallow the insecticide, accumulating and concentrating it in their bodies. Dr Barker found deposits of DDT throughout the digestive tracts of the worms, their blood vessels, nerves, and body wall. Undoubtedly some of the earthworms themselves succumb, but others survive to become 'biological magnifiers' of the poison. In the spring the robins return to provide another link in the cycle. As few as eleven large earthworms can transfer a lethal dose of DDT to a robin. And eleven worms form a small part of a day's rations to a bird that eats ten to twelve earthworms in as many minutes.

Not all robins receive a lethal dose, but another consequence may lead to the extinction of their kind as surely as fatal poisoning. The shadow of sterility lies over all the bird studies and indeed lengthens to include all living things within its potential range. There are now only two or three dozen robins to be found each spring on the entire 185-acre campus of Michigan State University, compared with a conservatively estimated 370 adults in this area before spraying. In 1954 every robin nest under observation by Mehner produced young. Towards the end of June, 1957, when at least 370 young birds (the normal replacement of the adult population) would have been foraging over the campus in the years before spraying began, Mehner could find *only one young robin.* A year later Dr Wallace was to report:

At no time during the spring or summer [of 1958] did I see a fledgling robin anywhere on the main campus, and so far I have failed to find anyone else who has seen one there.

Part of this failure to produce young is due, of course, to the fact that one or more of a pair of robins dies before the nesting cycle is completed. But Wallace has significant records which point to something more sinister – the actual destruction of the birds' capacity to reproduce. He has, for example,

records of robins and other birds building nests but laying no eggs, and others laying eggs and incubating them but not hatching them. We have one record of a robin that sat on its eggs faithfully for twenty-one days and they did not hatch. The normal incubation period is thirteen days.... Our analyses are showing high concentrations of DDT in the testes and ovaries of breeding birds [he told a congressional committee in 1960]. Ten males had amounts ranging from 30 to 109 parts per million in the testes, and two females had 151 and 211 parts per million respectively in the egg follicles in their ovaries.

Soon studies in other areas began to develop findings equally dismal. Professor Joseph Hickey and his students at the University of Wisconsin, after careful comparative studies of sprayed and unsprayed areas, reported the robin mortality to be at least 86 to 88 per cent. The Cranbrook Institute of Science at Bloomfield Hills, Michigan, in an effort to assess the extent of bird loss caused by the spraying of the elms, asked in 1956 that all birds thought to be victims of DDT poisoning be turned in to the institute for examination. The request had a response beyond all expectations. Within a few weeks the deep-freeze facilities of the institute were taxed to capacity, so that other specimens had to be refused. By 1959 a thousand poisoned birds from this single community had been turned in or reported. Although the robin was the chief victim (one woman calling the institute reported twelve robins lying dead on her lawn as she spoke), sixty-three different species were included among the specimens examined at the institute.

The robins, then, are only one part of the chain of devastation linked to the spraying of the elms, even as the elm programme is only one of the multitudinous spray programmes that cover our land with poisons. Heavy mortality has occurred among about ninety species of birds, including those most familiar to suburbanites and amateur naturalists. The populations of nesting birds in general have declined as much as 90 per cent in some of the sprayed towns. As we shall see, all the various types of birds are affected – ground feeders, tree-top feeders, bark feeders, predators.

It is only reasonable to suppose that all birds and mammals heavily dependent on earthworms or other soil organisms for food are threatened by the robins' fate. Some forty-five species of birds include earthworms in their diet. Among them is the woodcock, a species that winters in southern areas recently heavily sprayed with heptachlor. Two significant discoveries have now been made about the woodcock. Production of young birds on the New Brunswick breeding grounds is definitely reduced, and adult birds that have been analysed contain large residues of DDT and heptachlor.

Already there are disturbing records of heavy mortality among more than twenty other species of ground-feeding birds whose food – worms, ants, grubs, or other soil organisms – has been poisoned. These include three of the thrushes whose songs are among the most exquisite of bird voices, the olive-backed, the wood, and the hermit. And the sparrows that flit through the shrubby understory of the woodlands and forage with rustling sounds amid the fallen leaves – the song sparrow and the white-throat – these, too, have been found among the victims of the elm sprays.

Mammals, also, may easily be involved in the cycle, directly or indirectly. Earthworms are important among the various foods of the raccoon, and are eaten in the spring and autumn by opossums. Such subterranean tunnellers as shrews and moles capture them in some numbers, and then perhaps pass on the poison to predators such as screech owls and barn owls. Several dying screech owls were picked up in Wisconsin following heavy rains in spring, perhaps poisoned by feeding on earthworms. Hawks and owls have been found in convulsions – great horned owls, screech owls, red-shouldered hawks, sparrowhawks, marsh hawks. These may be cases of secondary poisoning, caused by eating birds or mice that have accumulated insecticides in their livers or other organs.

Nor is it only the creatures that forage on the ground or those who prey on them that are endangered by the foliar spraying of the elms. All of the tree-top feeders, the birds that glean their insect food from the leaves, have disappeared from heavily sprayed areas, among them those woodland sprites the kinglets, both ruby-crowned and golden-crowned, the tiny gnatcatchers, and many of the warblers, whose migrating hordes flow through the trees in spring in a multi-coloured tide of life. In 1956, a late spring delayed spraying so that it coincided with the arrival of an exceptionally heavy wave of warbler migration. Nearly all species of warblers present in the area were represented in the heavy kill that followed. In Whitefish Bay, Wisconsin, at least a thousand myrtle warblers could be seen in migration during former years; in 1958, after the spraying of the elms, observers could find only two. So, with additions from other communities, the list grows, and the warblers killed by the spray include those that most charm and fascinate all who are aware of them: the black-and-white, the yellow, the magnolia, and the Cape May; the oven-bird, whose call throbs in the May-time woods; the Blackburnian, whose wings are touched with flame; the chestnut-sided, the Canadian, and the black-throated green. These tree-top feeders are affected either directly by eating poisoned insects or indirectly by a shortage of food.

The loss of food has also struck hard at the swallows that cruise the skies, straining out the aerial insects as herring strain the plankton of the sea. A Wisconsin naturalist reported:

Swallows have been hard hit. Everyone complains of how few they have compared to four or five years ago. Our sky overhead was full of them only four years ago. Now we seldom see any.... This could be both lack of insects because of spray, or poisoned insects.

Of other birds this same observer wrote:

Another striking loss is the phoebe. Flycatchers are scarce everywhere but the early hardy common phoebe is no more. I've seen one this spring and only one last spring. Other birders in Wisconsin make the same complaint. I have had five or six pair of cardinals in the past, none now. Wrens, robins, catbirds and screech owls have nested each year in our garden. There are none now. Summer mornings are without bird song. Only pest birds, pigeons, starlings and English sparrows remain. It is tragic and I can't bear it.

The dormant sprays applied to the elms in the autumn, sending the poison into every little crevice in the bark, are probably responsible for the severe reduction observed in the number of chickadees, nuthatches, titmice, woodpeckers, and brown creepers. During the winter of 1957–8, Dr Wallace saw no chickadees or nuthatches at his home feeding station for the first time in many years. Three nuthatches he found later provided a sorry little step-by-step lesson in cause and effect: one was feeding on an elm, another was found dying of typical DDT symptoms, the third was dead. The dying nuthatch was later found to have 226 parts per million of DDT in its tissues.

The feeding habits of all these birds not only make them especially vulnerable to insect sprays but also make their loss a deplorable one for economic as well as less tangible reasons. The summer food of the white-breasted nuthatch and the brown creeper, for example, includes the eggs, larvae, and adults of a very large number of insects injurious to trees. About three-quarters of the food of the chickadee is animal, including all stages of the life cycle of many insects. The chickadee's method of feeding is described in Bent's monumental *Life Histories* of North American birds:

As the flock moves along each bird examines minutely bark, twigs, and branches, searching for tiny bits of food (spiders' eggs, cocoons, or other dormant insect life).

Various scientific studies have established the critical role of birds in insect control in various situations. Thus, woodpeckers are the primary control of the Engelmann spruce beetle, reducing its populations from 45 to 98 per cent, and are important in the control of the codling moth in apple orchards. Chickadees and other winter-resident birds can protect orchards against the cankerworm.

But what happens in nature is not allowed to happen in the modern, chemical-drenched world, where spraying destroys not only the insects but their principal enemy, the birds. When later there is a resurgence of the insect population, as almost always happens, the birds are not there to keep their numbers in check. As the Curator of Birds at the Milwaukee Public Museum, Owen J. Gromme, wrote to the Milwaukee *Journal*:

The greatest enemy of insect life is other predatory insects, birds, and some small mammals, but DDT kills indiscriminately, including nature's own safeguards or policemen.... In the name of progress are we to become victims of our own diabolical means of insect control to provide temporary comfort, only to lose out to destroying insects later on? By what means will we control new pests, which will attack remaining tree species after the elms are gone, when nature's safeguards (the birds) have been wiped out by poison?

Mr Gromme reported that calls and letters about dead and dying birds had been increasing steadily during the years since spraying began in Wisconsin. Questioning always revealed that spraying or fogging had been done in the area where the birds were dying.

Mr Gromme's experience has been shared by ornithologists and conservationists at most of the research centres of the Midwest such as the Cranbrook Institute in Michigan, the Illinois Natural History Survey, and the University of Wisconsin. A glance at the Letters-from-Readers column of newspapers almost anywhere that spraying is being done makes clear the fact that citizens are not only becoming aroused and indignant but that often they show a keener understanding of the dangers and inconsistencies of spraying than do the officials who order it to be done.

I am dreading the days to come soon now when many beautiful birds will be dying in our back yard [wrote a Milwaukee woman]. This is a pitiful, heartbreaking experience.... It is, moreover, frustrating and exasperating, for it evidently does not serve the purpose this slaughter was intended to serve.... Taking a long look, can you save trees without also saving birds? Do they not, in the economy of nature, save each other? Isn't it possible to help the balance of nature without destroying it?

The idea that the elms, majestic shade trees though they are, are not 'sacred cows' and do not justify an 'open end' campaign of destruction against all other forms of life is expressed in other letters. 'I have always loved our elm trees which seemed like trademarks on our landscape,' wrote another Wisconsin woman. 'But there are many kinds of trees.... We must save our birds, too. Can anyone imagine anything so cheerless and dreary as a springtime without a robin's song?'

To the public the choice may easily appear to be one of stark black-or-white simplicity: Shall we have birds or shall we have elms? But it is not as simple as that, and by one of the ironies that abound throughout the field of chemical control we may very well end by having neither if we continue on our present, well-travelled road. Spraying is killing the birds but it is not saving the elms. The illusion that salvation of the elms lies at the end of a spray nozzle is a dangerous will-o'-the-wisp that is leading one community after another into a morass of heavy expenditures, without producing lasting results. Greenwich, Connecticut, sprayed regularly for ten years. Then a drought year brought conditions especially favourable to the beetle and the mortality of elms went up 1,000 per cent. In Urbana, Illinois, where the University of Illinois is located, Dutch elm disease first appeared in 1951. Spraying was undertaken in 1953. By 1959, in spite of six years' spraying, the university campus had lost 86 per cent of its elms, half of them victims of Dutch elm disease.

In Toledo, Ohio, a similar experience caused the Superintendent of Forestry, Joseph A. Sweeney, to take a realistic look at the results of spraying. Spraying was begun there in 1953 and continued through 1959. Meanwhile, however, Mr Sweeney had noticed that a city-wide infestation of the cottony maple scale was worse after the spraying recommended by 'the books and the authorities' than it had been before. He decided to review the results of spraying for Dutch elm disease for himself. His findings shocked him. In the city of Toledo, he found,

the only areas under any control were the areas where we used some promptness in removing the diseased or brood trees. Where we depended on spraying the disease was out of control. In the country where nothing has been done the disease has not spread as fast as it has in the city. This indicates that spraying destroys any natural enemies.

We are abandoning spraying for the Dutch elm disease. This has brought me into conflict with the people who back any recommendations by the United States Department of Agriculture but I have the facts and will stick with them.

It is difficult to understand why these mid-western towns, to which the elm disease spread only rather recently, have so unquestioningly embarked on ambitious and expensive spraying programmes, apparently without waiting to inquire into the experience of other areas that have had longer acquaintance with the problem. New York State, for example, has certainly had the longest history of continuous experience with Dutch elm disease, for it was via the Port of New York that diseased elm wood is thought to have entered the United States about 1930. And New York State today has a most impressive record of containing and suppressing the disease. Yet it has not relied upon spraying. In fact, its agricultural extension service does not recommend spraying as a community method of control.

How, then, has New York achieved its fine record? From the early years of the battle for the elms to the present time, it has relied upon rigorous sanitation, or the prompt removal and destruction of all diseased or infected wood. In the beginning some of the results were disappointing, but this was because it was not at first understood that not only diseased trees but all elm wood in which the beetles might breed must be destroyed. Infected elm wood, after being cut and stored for firewood, will release a crop of fungus-carrying beetles unless burned before spring. It is the adult beetles, emerging from hibernation to feed in late April and May, that transmit Dutch elm disease. New York entomologists have learned by experience what kinds of beetle-breeding material have real importance in the spread of the disease. By concentrating on this dangerous material, it has been possible not only to get good results, but to keep the cost of the sanitation programme within reasonable limits. By 1950 the incidence of Dutch elm disease in New York City had been reduced to $\frac{2}{10}$ of 1 per cent of the city's 55,000 elms. A sanitation programme was launched in Westchester County in 1942. During the next fourteen years the average annual loss of elms was only $\frac{3}{10}$ of 1 per cent a year. Buffalo, with 185,000 elms, has an excellent record of containing the disease by sanitation, with recent annual losses amounting to only $\frac{3}{10}$ of 1 per cent. In other words, at this rate of loss it would take about 300 years to eliminate Buffalo's elms.

What has happened in Syracuse is especially impressive. There, no effective programme was in operation before 1957. Between 1951 and 1956 Syracuse lost nearly 3,000 elms. Then, under the direction of Howard C. Miller of the New York State University College of Forestry, an intensive drive was made to remove all diseased elm trees and all possible sources of beetle-breeding elm wood. The rate of loss is now well below 1 per cent a year.

The economy of the sanitation method is stressed by New York experts in Dutch elm disease control.

In most cases the actual expense is small compared with the probable saving [says J. G. Matthysse of the New York State College of Agriculture]. If it is a case of a dead or broken limb, the limb would have to be removed eventually, as a precaution against possible property damage or personal injury. If it is a fuel-wood pile, the wood can be used before spring, the bark can be peeled from the wood, or the wood can be stored in a dry place. In the case of dying or dead elm trees, the expense of prompt removal to prevent Dutch elm disease spread is usually no greater than would be necessary later, for most dead trees in urban regions must be removed eventually.

The situation with regard to Dutch elm disease is therefore not entirely hopeless provided informed and intelligent measures are taken. While it cannot be eradicated by any means now known, once it has become established in a community, it can be suppressed and contained within reasonable bounds by sanitation, and without the use of methods that are not only futile but involve tragic destruction of bird life. Other possibilities lie within the field of forest genetics, where experiments offer hope of developing a hybrid elm resistant to Dutch elm disease. The European elm is highly resistant, and many of them have been planted in Washington, D.C. Even during a period when a high percentage of the city's elms were affected, no cases of Dutch elm disease were found among these trees.

Replanting through an immediate tree nursery and forestry programme is being urged in communities that are losing large numbers of elms. This is important, and although such programmes might well include the resistant European elms, they should aim at a variety of species so that no future epidemic could deprive a community of its trees. The key to a healthy plant or animal community lies in what the British ecologist Charles Elton calls 'the conservation of variety'. What is happening now is in large part a result of the biological unsophistication of past generations. Even a generation ago no one knew that to fill large areas with a single species of tree was to invite disaster. And so whole towns lined their streets and dotted their parks with elms, and today the elms die and so do the birds.

Like the robin, another American bird seems to be on the verge of extinction. This is the national symbol, the eagle. Its populations have dwindled alarmingly within the past decade. The facts suggest that something is at work in the eagle's environment which has virtually destroyed its ability to reproduce. What this may be is not yet definitely known, but there is some evidence that insecticides are responsible.

The most intensively studied eagles in North America have been those nesting along a stretch of coast from Tampa to Fort Myers on the western coast of Florida. There a retired banker from Winnipeg, Charles Broley, achieved ornithological fame by banding more than 1,000 young bald eagles during the years 1939–49. (Only 166 eagles had been banded in all the earlier history of bird-banding.) Mr Broley banded eagles as young birds during the winter months before they had left their nests. Later recoveries of banded birds showed that these Florida-born eagles range northward along the coast into Canada as far as Prince Edward Island, although they had previously been considered non-migratory. In the autumn they return to the south, their migration being observed at such famous vantage points as Hawk Mountain in eastern Pennsylvania.

During the early years of his banding, Mr Broley used to find 125 active nests a year on the stretch of coast he had chosen for his work. The number of young banded each year was about 150. In 1947 the production of young birds began to decline. Some nests contained no eggs; others contained eggs that failed to hatch. Between 1952 and 1957, about 80 per cent of the nests failed to produce young. In the last year of this period only 43 nests were occupied. Seven of them produced young (8 eaglets); 23 contained eggs that failed to hatch; 13 were used merely as feeding stations by adult eagles and contained no eggs. In 1958 Mr Broley ranged over 100 miles of coast before finding and banding one eaglet. Adult eagles, which had been seen at 43 nests in 1957, were so scarce that he observed them at only 10 nests.

Although Mr Broley's death in 1959 terminated this valuable series of uninterrupted observations, reports by the Florida Audubon Society, as well as from New Jersey and Pennsylvania, confirm the trend that may well make it necessary for us to find a new national emblem. The reports of Maurice Broun, curator of the Hawk Mountain Sanctuary, are especially significant. Hawk Mountain is a picturesque mountain-top in south-eastern Pennsylvania, where the easternmost ridges of the Appalachians form a last barrier to the westerly winds before dropping away towards the coastal plain. Winds striking the mountains are deflected upward so that on many autumn days there is a continuous updraught on which the broad-winged hawks and eagles ride without effort, covering many miles of their southward migration in a day. At Hawk Mountain the ridges converge and so do the aerial highways. The result is that from a widespread territory to the north birds pass through this traffic bottleneck.

In his more than a score of years as custodian of the sanctuary there, Maurice Broun has observed and actually tabulated more hawks and eagles than any other American. The peak of the bald eagle migration comes in late August and early September. These are assumed to be Florida birds, returning to home territory after a summer in the North. (Later in the autumn and early winter a few larger eagles drift through. These are thought to belong to a northern race, bound for an unknown wintering ground.) During the first years after the sanctuary was established, from 1935 to 1939, 40 per cent of the eagles observed were yearlings, easily identified by their uniformly dark plumage. But in recent years these immature birds have become a rarity. Between 1955 and 1959, they made up only 20 per cent of the total count, and in one year (1957) there was only one young eagle for every thirty-two adults.

Observations at Hawk Mountain are in line with findings elsewhere. One such report comes from Elton Fawks, an official of the Natural Resources Council of Illinois. Eagles – probably northern nesters – winter along the Mississippi and Illinois Rivers. In 1958 Mr Fawks reported that a recent count of fifty-nine eagles had included only one immature bird. Similar indications of the dying out of the race come from the world's only sanctuary for eagles alone, Mount Johnson Island in the Susquehanna River. The island, although only eight miles above Conowingo Dam and about half a mile out from the Lancaster County shore, retains its primitive wildness. Since 1934 its single eagle nest has been under observation by Professor Herbert H.

Beck, an ornithologist of Lancaster and custodian of the sanctuary. Between 1935 and 1947 use of the nest was regular and
uniformly successful. Since 1947, although the adults have occupied the nest and there is evidence of egg-laying, no young
eagles have been produced.

On Mount Johnson Island as well as in Florida, then, the same
situation prevails – there is some occupancy of nests by adults,
some production of eggs, but few or no young birds. In seeking
an explanation, only one appears to fit all the facts. This is that
the reproductive capacity of the birds has been so lowered by
some environmental agent that there are now almost no annual
additions of young to maintain the race.

Exactly this sort of situation has been produced artificially in
other birds by various experimenters, notably Dr James DeWitt
of the United States Fish and Wildlife Service. Dr De Witt's now
classic experiments on the effect of a series of insecticides on
quail and pheasants have established the fact that exposure to
DDT or related chemicals, even when doing no observable harm
to the parent birds, may seriously affect reproduction. The way
the effect is exerted may vary, but the end result is always the
same. For example, quail into whose diet DDT was introduced
throughout the breeding season survived and even produced
normal numbers of fertile eggs. But few of the eggs hatched.
'Many embryos appeared to develop normally during the early
stages of incubation, but died during the hatching period,' Dr
DeWitt said. Of those that did hatch, more than half died within
five days. In other tests in which both pheasants and quail were
the subjects, the adults produced no eggs whatever if they had
been fed insecticide-contaminated diets throughout the year.
And at the University of California, Dr Robert Rudd and Dr
Richard Genelly reported similar findings. When pheasants
received dieldrin in their diets, 'egg production was markedly
lowered and chick survival was poor'. According to these authors, the delayed but lethal effect on the young birds follows
from storage of dieldrin in the yolk of the egg, from which it is
gradually assimilated during incubation and after hatching.

This suggestion is strongly supported by recent studies by
Dr Wallace and a graduate student, Richard F. Bernard, who
found high concentrations of DDT in robins on the Michigan

State University campus. They found the poison in all of the testes of male robins examined, in developing egg follicles, in the ovaries of females, in completed but unlaid eggs, in the oviducts, in unhatched eggs from deserted nests, in embryos within the eggs, and in a newly hatched, dead nestling.

These important studies establish the fact that the insecticidal poison affects a generation once removed from initial contact with it. Storage of poison in the egg, in the yolk material that nourishes the developing embryo, is a virtual death warrant and explains why so many of DeWitt's birds died in the egg or a few days after hatching.

Laboratory application of these studies to eagles presents difficulties that are nearly insuperable, but field studies are now under way in Florida, New Jersey, and elsewhere in the hope of acquiring definite evidence as to what has caused the apparent sterility of much of the eagle population. Meanwhile, the available circumstantial evidence points to insecticides. In localities where fish are abundant they make up a large part of the eagle's diet (about 65 per cent in Alaska; about 52 per cent in the Chesapeake Bay area). Almost unquestionably the eagles so long studied by Mr Broley were predominantly fish-eaters. Since 1945 this particular coastal area has been subjected to repeated sprayings with DDT dissolved in fuel oil. The principal target of the aerial spraying was the salt-marsh mosquito, which inhabits the marshes and coastal areas that are typical foraging areas for the eagles. Fishes and crabs were killed in enormous numbers. Laboratory analyses of their tissues revealed high concentrations of DDT – as much as 46 parts per million. Like the grebes of Clear Lake, which accumulated heavy concentrations of insecticide residues from eating the fish of the lake, the eagles have almost certainly been storing up the DDT in the tissues of their bodies. And like the grebes, the pheasants, the quail, and the robins, they are less and less able to produce young and to preserve the continuity of their race.

From all over the world come echoes of the peril that faces birds in our modern world. The reports differ in detail, but always repeat the theme of death to wildlife in the wake of pesticides. Such are the stories of hundreds of small birds and partridges dying in France after vine stumps were treated with an arsenic-containing herbicide, or of partridge shoots in Belgium, once famous for the numbers of their birds, denuded of partridges after the spraying of nearby farmlands.

In England the major problem seems to be a specialized one, linked with the growing practice of treating seed with insecticides before sowing. Seed treatment is not a wholly new thing, but in earlier years the chemicals principally used were fungicides. No effects on birds seem to have been noticed. Then about 1956 there was a change to dual-purpose treatment; in addition to a fungicide, dieldrin, aldrin, or heptachlor was added to combat soil insects. Thereupon the situation changed for the worse.

In the spring of 1960 a deluge of reports of dead birds reached British wildlife authorities, including the British Trust for Ornithology, the Royal Society for the Protection of Birds, and the Game Birds Association.

The place is like a battlefield [a landowner in Norfolk wrote]. My keeper has found innumerable corpses, including masses of small birds —chaffinches, greenfinches, linnets, hedge sparrows, also house sparrows...the destruction of wildlife is quite pitiful. [A gamekeeper wrote]: My partridges have been wiped out with the dressed corn, also some pheasants and all other birds, hundreds of birds have been killed.... As a lifelong gamekeeper it has been a distressing experience for me. It is bad to see pairs of partridges that have died together.

In a joint report, the British Trust for Ornithology and the Royal Society for the Protection of Birds described some 67 kills of birds – a far from complete listing of the destruction that took place in the spring of 1960. Of these 67, 59 were caused by seed dressings, 8 by toxic sprays.

A new wave of poisoning set in the following year. The death of 600 birds on a single estate in Norfolk was reported to the House of Lords, and 100 pheasants died on a farm in North Essex. It soon became evident that more counties were involved than in 1960 (34 compared with 23). Lincolnshire, heavily agricultural, seemed to have suffered most, with reports of 10,000 birds dead. But destruction involved all of agricultural Britain, from Angus in the north to Cornwall in the south, from Anglesey in the west to Norfolk in the east.

In the spring of 1961 concern reached such a peak that a

special committee of the House of Commons made an investigation of the matter, taking testimony from farmers, landowners, and representatives of the Ministry of Agriculture and of various governmental and non-governmental agencies concerned with wildlife.

'Pigeons are suddenly dropping out of the sky dead,' said one witness. 'You can drive a hundred or two hundred miles outside London and not see a single kestrel,' reported another. 'There has been no parallel in the present century, or at any time so far as I am aware, [this is] the biggest risk to wildlife and game that ever occurred in the country,' officials of the Nature Conservancy testified.

Facilities for chemical analysis of the victims were most inadequate to the task, with only two chemists in the country able to make the tests (one the government chemist, the other in the employ of the Royal Society for the Protection of Birds). Witnesses describe huge bonfires on which the bodies of the birds were burned. But efforts were made to have carcasses collected for examination, and of the birds analysed, all but one contained pesticide residues. The single exception was a snipe, which is not a seed-eating bird.

Along with the birds, foxes also may have been affected, probably indirectly by eating poisoned mice or birds. England, plagued by rabbits, sorely needs the fox as a predator. But between November 1959 and April 1960 at least 1,300 foxes died. Deaths were heaviest in the same counties from which sparrow hawks, kestrels, and other birds of prey virtually disappeared, suggesting that the poison was spreading through the food chain, reaching out from the seed-eaters to the furred and feathered carnivores. The actions of the moribund foxes were those of animals poisoned by chlorinated hydrocarbon insecticides. They were seen wandering in circles, dazed and half blind, before dying in convulsions.

The hearings convinced the committee that the threat to wildlife was 'most alarming'; it accordingly recommended to the House of Commons that

the Minister of Agriculture and the Secretary of State for Scotland should secure the immediate prohibition for the use as seed dressings of compounds containing dieldrin, aldrin, or heptachlor, or chemicals of comparable toxicity.

The committee also recommended more adequate controls to ensure that chemicals were adequately tested under field as well as laboratory conditions before being put on the market. This, it is worth emphasizing, is one of the great blank spots in pesticide research everywhere. Manufacturers' tests on the common laboratory animals – rats, dogs, guinea-pigs – include no wild species, no birds as a rule, no fishes, and are conducted under controlled and artificial conditions. Their application to wildlife in the field is anything but precise.

England is by no means alone in its problem of protecting birds from treated seeds. Here in the United States the problem has been most troublesome in the rice-growing areas of California and the South. For a number of years California ricegrowers have been treating seed with DDT as protection against tadpole shrimp and scavenger beetles which sometimes damage seedling rice. California sportsmen have enjoyed excellent hunting because of the concentrations of waterfowl and pheasants in the rice fields. But for the past decade persistent reports of bird losses, especially among pheasants, ducks, and blackbirds, have come from the rice-growing counties. 'Pheasant sickness' became a well-known phenomenon: birds 'seek water, become paralysed, and are found on the ditch banks and rice checks quivering', according to one observer. The 'sickness' comes in the spring, at the time the rice fields are seeded. The concentration of DDT used is many times the amount that will kill an adult pheasant.

The passage of a few years and the development of even more poisonous insecticides served to increase the hazard from treated seed. Aldrin, which is one hundred times as toxic as DDT to pheasants, is now widely used as a seed coating. In the rice fields of eastern Texas, this practice has seriously reduced the populations of the fulvous tree duck, a tawny-coloured, gooselike duck of the Gulf Coast. Indeed, there is some reason to think that the rice-growers, having found a way to reduce the populations of blackbirds, are using the insecticide for a dual purpose, with disastrous effects on several bird species of the rice fields.

As the habit of killing grows – the resort to 'eradicating' any creature that may annoy or inconvenience us – birds are more and more finding themselves a direct target of poisons rather than an incidental one. There is a growing trend towards aerial applications of such deadly poisons as parathion to 'control' concentrations of birds distasteful to farmers. The Fish and Wildlife Service has found it necessary to express serious concern over this trend, pointing out that 'parathion treated areas constitute a potential hazard to humans, domestic animals, and wildlife'. In southern Indiana, for example, a group of farmers went together in the summer of 1959 to engage a spray plane to treat an area of river bottomland with parathion. The area was a favoured roosting site for thousands of blackbirds that were feeding in nearby cornfields. The problem could have been solved easily by a slight change in agricultural practice – a shift to a variety of corn with deep-set ears not accessible to the birds – but the farmers had been persuaded of the merits of killing by poison, and so they sent in the planes on their mission of death.

The results probably gratified the farmers, for the casualty list included some 65,000 red-winged blackbirds and starlings. What other wildlife deaths may have gone unnoticed and unrecorded is not known. Parathion is not a specific for blackbirds: it is a universal killer. But such rabbits or raccoons or opossums as may have roamed those bottomlands and perhaps never visited the farmers' cornfields were doomed by a judge and jury who neither knew of their existence nor cared.

And what of human beings? In Californian orchards sprayed with this same parathion, workers handling foliage that had been treated *a month* earlier collapsed and went into shock, and escaped death only through skilled medical attention. Does Indiana still raise any boys who roam through woods or fields and might even explore the margins of a river? If so, who guarded the poisoned area to keep out any who might wander in, in misguided search for unspoiled nature? Who kept vigilant watch to tell the innocent stroller that the fields he was about to enter were deadly – all their vegetation coated with a lethal film? Yet at so fearful a risk the farmers, with none to hinder them, waged their needless war on blackbirds.

In each of these situations, one turns away to ponder the question: Who has made the decision that sets in motion these chains of poisonings, this ever-widening wave of death that spreads out, like ripples when a pebble is dropped into a still pond? Who has placed in one pan of the scales the leaves that might have been eaten by the beetles and in the other the pitiful heaps of many-hued feathers, the lifeless remains of the birds that fell before the unselective bludgeon of insecticidal poisons? Who has decided – who has the *right* to decide – for the countless legions of people who were not consulted that the supreme value is a world without insects, even though it be also a sterile world ungraced by the curving wing of a bird in flight? The decision is that of the authoritarian temporarily entrusted with power; he has made it during a moment of inattention by millions to whom beauty and the ordered world of nature still have a meaning that is deep and imperative.

Chapter 9

Rivers of Death

From the green depths of the offshore Atlantic many paths
lead back to the coast. They are paths followed by fish; although
unseen and intangible, they are linked with the outflow of waters
from the coastal rivers. For thousands upon thousands of years
the salmon have known and followed these threads of fresh water
that lead them back to the rivers, each returning to the tributary
in which it spent the first months or years of life. So, in the
summer and autumn of 1953, the salmon of the river called
Miramichi on the coast of New Brunswick moved in from their
feeding grounds in the far Atlantic and ascended their native
river. In the upper reaches of the Miramichi, in streams that
gather together a network of shadowed brooks, the salmon
deposited their eggs that autumn in beds of gravel over which
the stream water flowed swift and cold. Such places, the water-
sheds of the great coniferous forests of spruce and balsam, of
hemlock and pine, provide the kind of spawning grounds that
salmon must have in order to survive.

These events repeated a pattern that was age-old, a pattern
that had made the Miramichi one of the finest salmon streams in
North America. But that year the pattern was to be broken.

During the autumn and winter the salmon eggs, large and
thick-shelled, lay in shallow gravel-filled troughs, or redds,
which the mother fish had dug in the stream bottom. In the cold
of winter they developed slowly, as was their way, and only
when spring at last brought thawing and release to the forest
streams did the young hatch. At first they hid among the pebbles
of the stream bed – tiny fish about half an inch long. They
took no food, living on the large yolk sac. Not until it was
absorbed would they begin to search the stream for small
insects.

With the newly hatched salmon in the Miramichi that spring

of 1954 were young of previous hatchings, salmon a year or two old, young fish in brilliant coats marked with bars and bright red spots. These young fed voraciously, seeking out the strange and varied insect life of the stream.

As the summer approached, all this was changed. That year the watershed of the north-west Miramichi was included in a vast spraying programme which the Canadian Government had embarked upon the previous year – a programme designed to save the forests from the spruce budworm. The budworm is a native insect that attacks several kinds of evergreens. In eastern Canada it seems to become extraordinarily abundant about every thirty-five years. The early 1950s had seen such an upsurge in the budworm populations. To combat it, spraying with DDT was begun, first in a small way, then at a suddenly accelerated rate in 1953. Millions of acres of forests were sprayed instead of thousands as before, in an effort to save the balsams, which are the mainstay of the pulp and paper industry.

So in 1954, in the month of June, the planes visited the forests of the north-west Miramichi and white clouds of settling mist marked the crisscross pattern of their flight. The spray – one half pound of DDT to the acre in a solution of oil – filtered down through the balsam forests and some of it finally reached the ground and the flowing streams. The pilots, their thoughts only on their assigned task, made no effort to avoid the streams or to shut off the spray nozzles while flying over them; but because spray drifts so far in even the slightest stirrings of air, perhaps the result would have been little different if they had.

Soon after the spraying had ended there were unmistakable signs that all was not well. Within two days dead and dying fish, including many young salmon, were found along the banks of the stream. Brook trout also appeared among the dead fish, and along the roads and in the woods birds were dying. All the life of the stream was stilled. Before the spraying there had been a rich assortment of the water life that forms the food of salmon and trout – caddis fly larvae, living in loosely fitting protective cases of leaves, stems or gravel cemented together with saliva, stonefly nymphs clinging to rocks in the swirling currents, and the wormlike larvae of blackflies edging the stones under riffles or where the stream spills over steeply slanting rocks. But now the stream insects were dead, killed by the DDT, and there was nothing for a young salmon to eat.

Amid such a picture of death and destruction, the young salmon themselves could hardly have been expected to escape, and they did not. By August not one of the young salmon that had emerged from the gravel beds that spring remained. A whole year's spawning had come to nothing. The older young, those hatched a year or more earlier, fared only slightly better. For every six young of the 1953 hatch that had foraged in the stream as the planes approached, only one remained. Young salmon of the 1952 hatch, almost ready to go to sea, lost a third of their numbers.

All these facts are known because the Fisheries Research Board of Canada had been conducting a salmon study on the north-west Miramichi since 1950. Each year it had made a census of the fish living in this stream. The records of the biologists covered the number of adult salmon ascending to spawn, the number of young of each age group present in the stream, and the normal population not only of salmon but of other species of fish inhabiting the stream. With this complete record of pre-spraying conditions, it was possible to measure the damage done by the spraying with an accuracy that has seldom been matched elsewhere.

The survey showed more than the loss of young fish; it revealed a serious change in the streams themselves. Repeated sprayings have now completely altered the stream environment, and the aquatic insects that are the food of salmon and trout have been killed. A great deal of time is required, even after a single spraying, for most of these insects to build up sufficient numbers to support a normal salmon population – time measured in years rather than months.

The smaller species, such as midges and blackflies, become re-established rather quickly. These are suitable food for the smallest salmon, the fry only a few months old. But there is no such rapid recovery of the larger aquatic insects, on which salmon in their second and third years depend. These are the larval stages of caddis flies, stoneflies, and mayflies. Even in the second year after DDT enters a stream, a foraging salmon parr would have trouble finding anything more than an occasional small stonefly.

There would be no large stoneflies, no mayflies, no caddis flies. In an effort to supply this natural food, the Canadians have attempted to transplant caddis fly larvae and other insects to the barren reaches of the Miramichi. But of course such transplants would be wiped out by any repeated spraying.

The budworm populations, instead of dwindling as expected, have proved refractory, and from 1955 to 1957 spraying was repeated in various parts of New Brunswick and Quebec, some places being sprayed as many as three times. By 1957, nearly fifteen million acres had been sprayed. Although spraying was then tentatively suspended, a sudden resurgence of budworms led to its resumption in 1960 and 1961. Indeed, there is no evidence anywhere that chemical spraying for budworm control is more than a stop-gap measure (aimed at saving the trees from death through defoliation over several successive years), and so its unfortunate side-effects will continue to be felt as spraying is continued. In an effort to minimize the destruction of fish, the Canadian forestry officials have reduced the concentration of DDT from the half-pound previously used to a quarter-pound to the acre, on the recommendation of the Fisheries Research Board. (In the United States the standard and highly lethal pound-to-the-acre still prevails.) Now, after several years in which to observe the effects of spraying, the Canadians find a mixed situation, but one that affords very little comfort to devotees of salmon fishing, provided spraying is continued.

A very unusual combination of circumstances has so far saved the runs of the north-west Miramichi from the destruction that was anticipated – a constellation of happenings that might not occur again in a century. It is important to understand what has happened there, and the reasons for it.

In 1954, as we have seen, the watershed of this branch of the Miramichi was heavily sprayed. Thereafter, except for a narrow band sprayed in 1956, the whole upper watershed of this branch was excluded from the spraying programme. In the autumn of 1954 a tropical storm played its part in the fortunes of the Miramichi salmon. Hurricane Edna, a violent storm to the very end of its northward path, brought torrential rains to the New England and Canadian coasts. The resulting freshets carried streams of fresh water far out to sea and drew in unusual numbers of salmon. As a result, the gravel beds of the streams which the salmon seek out for spawning received an unusual abundance of eggs. The young salmon hatching in the north-west Miramichi in the spring of 1955 found circumstances practically ideal for their survival. While the DDT had killed off all stream insects the year before, the smallest of the insects – the midges and blackflies – had returned in numbers. These are the normal food of baby salmon. The salmon fry of that year not only found abundant food but they had few competitors for it. This was because of the grim fact that the older young salmon had been killed off by the spraying in 1954. Accordingly, the fry of 1955 grew very fast and survived in exceptional numbers. They completed their stream growth rapidly and went to sea early. Many of them returned in 1959 to give large runs of grilse to the native stream.

If the runs in the north-west Miramichi are still in relatively good condition this is because spraying was done in one year only. The results of repeated spraying are clearly seen in other streams of the watershed, where alarming declines in the salmon populations are occurring.

In all sprayed streams, young salmon of every size are scarce. The youngest are often 'practically wiped out', the biologists report. In the main south-west Miramichi, which was sprayed in 1956 and 1957, the 1959 catch was the lowest in a decade. Fishermen remarked on the extreme scarcity of grilse – the youngest group of returning fish. At the sampling trap in the estuary of the Miramichi the count of grilse was only a fourth as large in 1959 as the year before. In 1959 the whole Miramichi watershed produced only about 600,000 smolt (young salmon descending to the sea). This was less than a third of the runs of the three preceding years.

Against such a background, the future of the salmon fisheries in New Brunswick may well depend on finding a substitute for drenching forests with DDT.

The eastern Canadian situation is not unique, except perhaps in the extent of forest spraying and the wealth of facts that have been collected. Maine, too, has its forests of spruce and balsam, and its problem of controlling forest insects. Maine, too, has its salmon runs – a remnant of the magnificent runs of former days, but a remnant hard won by the work of biologists and conservationists to save some habitat for salmon in streams burdened with industrial pollution and choked with logs. Although spraying has been tried as a weapon against the ubiquitous budworm, the areas affected have been relatively small and have not, as yet, included important spawning streams for salmon. But what happened to stream fish in an area observed by the Maine Department of Inland Fisheries and Game is perhaps a portent of things to come.

Immediately after the 1958 spraying [the Department reported], moribund suckers were observed in large numbers in Big Goddard Brook. These fish exhibited the typical symptoms of DDT poisoning; they swam erratically, gasped at the surface, and exhibited tremors and spasms. In the first five days after spraying, 668 dead suckers were collected from two blocking nets. Minnows and suckers were also killed in large numbers in Little Goddard, Carry, Alder, and Blake Brooks. Fish were often seen floating passively downstream in a weakened and moribund condition. In several instances, blind and dying trout were found floating passively downstream more than a week after spraying.

(The fact that DDT may cause blindness in fish is confirmed by various studies. A Canadian biologist who observed spraying on northern Vancouver Island in 1957 reported that cut-throat trout fingerlings could be picked out of the streams by hand, for the fish were moving sluggishly and made no attempt to escape. On examination, they were found to have an opaque white film covering the eye, indicating that vision had been impaired or destroyed. Laboratory studies by the Canadian Department of Fisheries showed that almost all fish [Coho salmon] not actually killed by exposure to low concentrations of DDT [3 parts per million] showed symptoms of blindness, with marked opacity of the lens.)

Wherever there are great forests, modern methods of insect control threaten the fishes inhabiting the streams in the shelter of the trees. One of the best-known examples of fish destruction in the United States took place in 1955, as a result of spraying in and near Yellowstone National Park. By the autumn of that year, so many dead fish had been found in the Yellowstone

River that sportsmen and Montana fish-and-game administrators became alarmed. About ninety miles of the river were affected. In one 300-yard length of shoreline, 600 dead fish were counted, including brown trout, whitefish, and suckers. Stream insects, the natural food of trout, had disappeared.

Forest Service officials declared they had acted on advice that one pound of DDT to the acre was 'safe'. But the results of the spraying should have been enough to convince anyone that the advice had been far from sound. A cooperative study was begun in 1956 by the Montana Fish and Game Department and two federal agencies, the Fish and Wildlife Service and the Forest Service. Spraying in Montana that year covered 900,000 acres; 800,000 acres were also treated in 1957. The biologists therefore had no trouble finding areas for their study.

Always, the pattern of death assumed a characteristic shape: the smell of DDT over the forests, an oil film on the water surface, dead trout along the shoreline. All fish analysed, whether taken alive or dead, had stored DDT in their tissues. As in eastern Canada, one of the most serious effects of spraying was the severe reduction of food organisms. On many study areas aquatic insects and other stream-bottom fauna were reduced to a tenth of their normal populations. Once destroyed, populations of these insects, so essential to the survival of trout, take a long time to rebuild. Even by the end of the second summer after spraying, only meagre quantities of aquatic insects had re-established themselves, and on one stream – formerly rich in bottom fauna – scarcely any could be found. In this particular stream, game fish had been reduced by 80 per cent.

The fish do not necessarily die immediately. In fact, delayed mortality may be more extensive than the immediate kill and, as the Montana biologists discovered, it may go unreported because it occurs after the fishing season. Many deaths occurred in the study streams among autumn spawning fish, including brown trout, brook trout, and whitefish. This is not surprising, because in time of physiological stress the organism, be it fish or man, draws on stored fat for energy. This exposes it to the full lethal effect of the DDT stored in the tissues.

It was therefore more than clear that spraying at the rate of a pound of DDT to the acre posed a serious threat to the fishes in forest streams. Moreover, control of the budworm had not been achieved and many areas were scheduled for respraying. The Montana Fish and Game Department registered strong opposition to further spraying, saying it was 'not willing to compromise the sport fishery resource for programmes of questionable necessity and doubtful success'. The Department declared, however, that it would continue to cooperate with the Forest Service, 'in determining ways to minimize adverse effects'.

But can such cooperation actually succeed in saving the fish? An experience in British Columbia speaks volumes on this point. There an outbreak of the black-headed budworm had been raging for several years. Forestry officials, fearing that another season's defoliation might result in severe loss of trees, decided to carry out control operations in 1957. There were many consultations with the Game Department, whose officials were concerned about the salmon runs. The Forest Biology Division agreed to modify the spraying programme in every possible way short of destroying its effectiveness, in order to reduce risks to the fish.

Despite these precautions, and despite the fact that a sincere effort was apparently made, *in at least four major streams almost 100 per cent of the salmon were killed.*

In one of the rivers, the young of a run of 40,000 adult Coho salmon were almost completely annihilated. So were the young stages of several thousand steelhead trout and other species of trout. The Coho salmon has a three-year life cycle and the runs are composed almost entirely of fish of a single age group. Like other species of salmon, the Coho has a strong homing instinct, returning to its natal stream. There will be no repopulation from other streams. This means, then, that every third year the run of salmon into this river will be almost non-existent, until such time as careful management, by artificial propagation or other means, has been able to rebuild this commercially important run.

There are ways to solve this problem – to preserve the forests and to save the fishes, too. To assume that we must resign ourselves to turning our waterways into rivers of death is to follow the counsel of despair and defeatism. We must make wider use of alternative methods that are now known, and we must devote our ingenuity and resources to developing others. There are cases on record where natural parasitism has kept the budworm under control more effectively than spraying. Such natural control needs to be utilized to the fullest extent. There are possibilities of using less toxic sprays or, better still, of introducing microorganisms that will cause disease among the budworms without affecting the whole web of forest life. We shall see later what some of these alternative methods are and what they promise. Meanwhile, it is important to realize that chemical spraying of forest insects is neither the only way nor the best way.

The pesticide threat to fishes may be divided into three parts. One, as we have seen, relates to the fishes of running streams in northern forests and to the single problem of forest spraying. It is confined almost entirely to the effects of DDT. Another is vast, sprawling, and diffuse, for it concerns the many different kinds of fishes – bass, sunfish, crappies, suckers, and others – that inhabit many kinds of waters, still or flowing, in many parts of the country. It also concerns almost the whole gamut of insecticides now in agricultural use, although a few principal offenders like endrin, toxaphene, dieldrin, and heptachlor can easily be picked out. Still another problem must now be considered largely in terms of what we may logically suppose will happen in the future, because the studies that will disclose the facts are only beginning to be made. This has to do with the fishes of salt marshes, bays, and estuaries.

It was inevitable that serious destruction of fishes would follow the widespread use of the new organic pesticides. Fishes are almost fantastically sensitive to the chlorinated hydrocarbons that make up the bulk of modern insecticides. And when millions of tons of poisonous chemaicls are applied to the surface of the land, it is inevitable that some of them will find their way into the ceaseless cycle of waters moving between land and sea.

Reports of fish kills, some of disastrous proportions, have now become so common that the United States Public Health Service has set up an office to collect such reports from the states as an index of water pollution.

This is a problem that concerns a great many people. Some twenty-five million Americans look to fishing as a major source of recreation and another fifteen million are at least casual anglers. These people spend three billion dollars annually for licences, tackle, boats, camping equipment, petrol, and lodgings. Anything that deprives them of their sport will also reach out and affect a large number of economic interests. The commercial fisheries represent such an interest, and even more importantly, an essential source of food. Inland and coastal fisheries (excluding the offshore catch) yield an estimated three billion pounds a year. Yet, as we shall see, the invasion of streams, ponds, rivers, and bays by pesticides is now a threat to both recreational and commercial fishing.

Examples of the destruction of fish by agricultural crop sprayings and dustings are everywhere to be found. In California, for example, the loss of some 60,000 game fish, mostly bluegill and other sunfish, followed an attempt to control the rice-leaf miner with dieldrin. In Louisiana thirty or more instances of heavy fish mortality occurred in one year alone (1960) because of the use of endrin in the sugar-cane fields. In Pennsylvania fish have been killed in numbers by endrin, used in orchards to combat mice. The use of chlordane for grasshopper control on the high western plains has been followed by the death of many stream fish.

Probably no other agricultural programme has been carried out on so large a scale as the dusting and spraying of millions of acres of land in southern United States to control the fire ant. Heptachlor, the chemical chiefly used, is only slightly less toxic to fish than DDT. Dieldrin, another fire ant poison, has a well-documented history of extreme hazard to all aquatic life. Only endrin and toxaphene represent a greater danger to fish.

All areas within the fire ant control area, whether treated with heptachlor or dieldrin, reported disastrous effects on aquatic life. A few excerpts will give the flavour of the reports from biologists who studied the damage: From Texas, 'Heavy loss of aquatic life despite efforts to protect canals', 'Dead fish... were present in all treated water', 'Fish kill was heavy and continued for over three weeks.' From Alabama, 'Most adult fish were killed [in Wilcox County] within a few days after treatment', 'The fish in temporary waters and small tributary streams appeared to have been completely eradicated.'

In Louisiana, farmers complained of loss in farm ponds.

Along one canal bank more than 500 dead fish were seen floating or lying on the bank on a stretch of less than a quarter of a mile. In another parish 150 dead sunfish could be found for every four that remained alive. Five other species appeared to have been wiped out completely.

In Florida, fish from ponds in a treated area were found to contain residues of heptachlor and a derived chemical, heptachlor epoxide. Included among these fish were sunfish and bass, which of course are favourites of anglers and commonly find their way to the dinner-table. Yet the chemicals they contained are among those the Food and Drug Administration considers too dangerous for human consumption, even in minute quantities.

So extensive were the reported kills of fish, frogs, and other life of the waters that the American Society of Ichthyologists and Herpetologists, a venerable scientific organization devoted to the study of fishes, reptiles, and amphibians, passed a resolution in 1958 calling on the Department of Agriculture and the associated state agencies to cease 'aerial distribution of heptachlor, dieldrin, and equivalent poisons – before irreparable harm is done'. The Society called attention to the great variety of species of fish and other forms of life inhabiting the south-eastern part of the United States, including species that occur nowhere else in the world. 'Many of these animals,' the Society warned, 'occupy only small areas and therefore might readily be completely exterminated.'

Fishes of the southern states have also suffered heavily from insecticides used against cotton insects. The summer of 1950 was a season of disaster in the cotton-growing country of northern Alabama. Before that year, only limited use had been made of organic insecticides for the control of the boll weevil. But in 1950 there were many weevils because of a series of mild winters, and so an estimated 80 to 95 per cent of the farmers, on the urging of the county agents, turned to the use of insecticides. The chemical most popular with the farmers was toxaphene, one of the most destructive to fishes.

Rains were frequent and heavy that summer. They washed the chemicals into the streams, and as this happened the farmers applied more. An average acre of cotton that year received 63 pounds of toxaphene. Some farmers used as much as 200 pounds per acre; one, in an extraordinary excess of zeal, applied more than a quarter of a ton to the acre.

The results could easily have been foreseen. What happened in Flint Creek, flowing through fifty miles of Alabama cotton country before emptying into Wheeler Reservoir, was typical of the region. On 1 August, torrents of rain descended on the Flint Creek watershed. In trickles, in rivulets, and finally in floods the water poured off the land into the streams. The water level rose six inches in Flint Creek. By the next morning it was obvious that a great deal more than rain had been carried into the stream. Fish swam about in aimless circles near the surface. Sometimes one would throw itself out of the water on to the bank. They could easily be caught; one farmer picked up several and took them to a spring-fed pool. There, in the pure water, these few recovered. But in the stream dead fish floated down all day. This was but the prelude to more, for each rain washed more of the insecticide into the river, killing more fish. The rain of 10 August resulted in such a heavy fish kill throughout the river that few remained to become victims of the next surge of poison into the stream, which occurred on 15 August. But evidence of the deadly presence of the chemicals was obtained by placing test goldfish in cages in the river; they were dead within a day.

The doomed fish of Flint Creek included large numbers of white crappies, a favourite among anglers. Dead bass and sunfish were also found, occurring abundantly in Wheeler Reservoir, into which the creek flows. All the rough-fish population of these waters was destroyed also – the carp, buffalo, drum, gizzard shad, and catfish. None showed signs of disease – only the erratic movements of the dying and a strange deep wine colour of the gills.

In the warm enclosed waters of farm ponds, conditions are very likely to be lethal for fish when insecticides are applied in the vicinity. As many examples show, the poison is carried in by rains and run-off from surrounding lands. Sometimes the ponds receive not only contaminated run-off but also a direct dose as crop-dusting pilots neglect to shut off the duster in passing over a pond. Even without such complications, normal agricultural use subjects fish to far heavier concentrations of chemicals than would be required to kill them. In other words, a marked reduction in the poundages used would hardly alter the lethal situation, for applications of over 0·1 pound per acre to the pond itself are generally considered hazardous. And the poison, once introduced, is hard to get rid of. One pond that had been treated with DDT to remove unwanted shiners remained so poisonous through repeated drainings and flushings that it killed 94 per cent of the sunfish with which it was later stocked. Apparently the chemical remained in the mud of the pond bottom.

Conditions are evidently no better now than when the modern insecticides first came into use. The Oklahoma Wildlife Conservation Department stated in 1961 that reports of fish losses in farm ponds and small lakes had been coming in at the rate of at least one a week, and that such reports were increasing. The conditions usually responsible for these losses in Oklahoma were those made familiar by repetition over the years: the application of insecticides to crops, a heavy rain, and poison washed into the ponds.

In some parts of the world the cultivation of fish in ponds provides an indispensable source of food. In such places the use of insecticides without regard for the effects on fish creates immediate problems. In Rhodesia, for example, the young of an important food fish, the Kafue bream, are killed by exposure to only 0·04 parts per million of DDT in shallow pools. Even smaller doses of many other insecticides would be lethal. The shallow waters in which these fish live are favourable mosquito-breeding places. The problem of controlling mosquitoes and at the same time conserving a fish important in the Central African diet has obviously not been solved satisfactorily.

Milkfish farming in the Philippines, China, Vietnam, Thailand, Indonesia, and India faces a similar problem. The milkfish is cultivated in shallow ponds along the coasts of these countries. Schools of young suddenly appear in the coastal waters (from no one knows where) and are scooped up and placed in impoundments, where they complete their growth. So important is this fish as a source of animal protein for the rice-eating millions of South East Asia and India that the Pacific Science Congress has recommended an international effort to search for the now unknown spawning grounds, in order to develop the farming of these fish on a massive scale. Yet spraying has been permitted to cause heavy losses in existing impoundments. In the Philippines aerial spraying for mosquito control has cost pond owners dearly. In one such pond containing 120,000 milkfish, more than half the fish died after a spray plane had passed over, in spite of desperate efforts by the owner to dilute the poison by flooding the pond.

One of the most spectacular fish kills of recent years occurred in the Colorado River below Austin, Texas, in 1961. Shortly after daylight on Sunday morning, 25 January, dead fish appeared in the new Town Lake in Austin and in the river for a distance of about five miles below the lake. None had been seen the day before. On Monday there were reports of dead fish fifty miles downstream. By this time it was clear that a wave of some poisonous substance was moving down in the river water. By 31 January, fish were being killed 100 miles downstream near La Grange, and a week later the chemicals were doing their lethal work 200 miles below Austin. During the last week of January the locks of the Intracoastal Waterway were closed to exclude the toxic waters from Matagorda Bay and divert them into the Gulf of Mexico.

Meanwhile, investigators in Austin noticed an odour associated with the insecticides chlordane and toxaphene. It was especially strong in the discharge from one of the storm sewers. This sewer had in the past been associated with trouble from industrial wastes, and when officers of the Texas Game and Fish Commission followed it back from the lake, they noticed an odour like that of benzene hexachloride at all openings as far back as a feeder line from a chemical plant. Among the major products of this plant were DDT, benzene hexachloride, chlordane, and toxaphene, as well as smaller quantities of other insecticides. The manager of the plant admitted that quantities of powdered insecticide had been washed into the storm sewer recently and, more significantly, he acknowledged that such disposal of insecticide spillage and residues had been common practice for the past ten years.

On searching further, the fishery officers found other plants where rains or ordinary clean-up waters would carry insecticides into the sewer. The fact that provided the final link in the chain, however, was the discovery that a few days before the water in lake and river became lethal to fish the entire storm-sewer system had been flushed out with several million gallons of water under high pressure to clear it of debris. This flushing had undoubtedly released insecticides lodged in the accumulation of gravel, sand, and rubble and carried them into the lake and thence to the river, where chemical tests later established their presence.

As the lethal mass drifted down the Colorado it carried death before it. For 140 miles downstream from the lake the kill of fish must have been almost complete, for when seines were used later in an effort to discover whether any fish had escaped they came up empty. Dead fish of twenty-seven species were observed, totalling about 1,000 pounds to a mile of river bank. There were channel cats, the chief game fish of the river. There were blue and flathead catfish, bullheads, four species of sunfish, shiners, dace, stone rollers, largemouth bass, carp, mullet, suckers. There were eels, gar, carp, river carpsuckers, gizzard shad, and buffalo. Among them were some of the patriarchs of the river, fish that by their size must have been of great age – many flathead catfish, weighing over 25 pounds, some of 60 pounds reportedly picked up by local residents along the river, and a giant blue catfish officially recorded as weighing 84 pounds.

The Game and Fish Commission predicted that even without further pollution the pattern of the fish population of the river would be altered for years. Some species – those existing at the limits of their natural range – might never be able to re-establish themselves, and the others could do so only with the aid of extensive stocking operations by the state.

This much of the Austin fish disaster is known, but there was almost certainly a sequel. The toxic river water was still possessed of its death-dealing power after passing more than 200 miles downstream. It was regarded as too dangerous to be admitted to the waters of Matagorda Bay, with its oyster beds and shrimp fisheries, and so the whole toxic outflow was diverted to the waters of the open Gulf. What were its effects there? And what of the outflow of scores of other rivers, carrying contaminants perhaps equally lethal?

At present our answers to these questions are for the most part only conjectures, but there is growing concern about the role of pesticide pollution in estuaries, salt marshes, bays, and other coastal waters. Not only do these areas receive the contaminated discharge of rivers but all too commonly they are sprayed directly in efforts to control mosquitoes or other insects.

Nowhere has the effect of pesticides on the life of salt marshes, estuaries, and all quiet inlets from the sea been more graphically demonstrated than on the eastern coast of Florida, in the Indian River country. There, in the spring of 1955, some 2,000 acres of salt marsh in St Lucie County were treated with dieldrin in an attempt to eliminate the larvae of the sandfly. The concentration used was one pound of active ingredient to the acre. The effect on the life of the waters was catastrophic. Scientists from the Entomology Research Center of the State Board of Health surveyed the carnage after the spraying and reported that the fish kill was 'substantially complete'. Everywhere dead fishes littered the shores. From the air sharks could be seen moving in, attracted by the helpless and dying fishes in the water. No species was spared. Among the dead were mullets, snook, mojarras, gambusia.

The minimum immediate over-all kill throughout the marshes, exclusive of the Indian River shoreline, was 20–30 tons of fishes, or about 1,175,000 fishes, of at least 30 species [reported R. W. Harrington, Jr, and W. L. Bidlingmayer of the survey team].

Molluscs seemed to be unharmed by dieldrin. Crustaceans were virtually exterminated throughout the area. The entire aquatic crab population was apparently destroyed and the fiddler crabs, all but annihilated, survived temporarily only in patches of marsh evidently missed by the pellets.

The larger game and food fishes succumbed most rapidly.... Crabs set upon and destroyed the moribund fishes, but the next day were dead themselves. Snails continued to devour fish carcasses. After two weeks, no trace remained of the litter of dead fishes.

The same melancholy picture was painted by the late Dr Herbert R. Mills from his observations in Tampa Bay on the opposite coast of Florida, where the National Audubon Society operates a sanctuary for seabirds in the area including Whiskey Stump Key. The sanctuary ironically became a poor refuge after the local health authorities undertook a campaign to wipe out the salt-marsh mosquitoes. Again fishes and crabs were the principal victims. The fiddler crab, that small and picturesque crustacean whose hordes move over mud flats or sand flats like grazing cattle, has no defence against the sprayers. After successive spraying during the summer and autumn months (some areas were sprayed as many as sixteen times), the state of the fiddler crabs was summed up by Dr Mills:

A progressive scarcity of fiddlers had by this time become apparent. Where there should have been in the neighbourhood of 100,000 fiddlers under the tide and weather conditions of the day [October 12] there were not over a hundred which could be seen anywhere on the beach, and these were all dead or sick, quivering, twitching, stumbling, scarcely able to crawl; although in neighbouring unsprayed areas fiddlers were plentiful.

The place of the fiddler crab in the ecology of the world it inhabits is a necessary one, not easily filled. It is an important source of food for many animals. Coastal raccoons feed on them. So do marsh-inhabiting birds like the clapper rail, shore-birds, and even visiting seabirds. In one New Jersey salt marsh sprayed with DDT, the normal population of laughing gulls was decreased by 85 per cent for several weeks, presumably because the birds could not find sufficient food after the spraying. The marsh fiddlers are important in other ways as well, being useful scavengers and aerating the mud of the marshes by their extensive burrowings. They also furnish quantities of bait for fishermen.

The fiddler crab is not the only creature of tidal marsh and estuary to be threatened by pesticides; others of more obvious importance to man are endangered. The famous blue crab of the Chesapeake Bay and other Atlantic Coast areas is an example. These crabs are so highly susceptible to insecticides that every spraying of creeks, ditches, and ponds in tidal marshes kills most of the crabs living there. Not only do the local crabs die, but others moving into a sprayed area from the sea succumb to the lingering poison. And sometimes poisoning may be indirect, as in the marshes near Indian River, where scavenger crabs attacked the dying fishes, but soon themselves succumbed to the poison. Less is known about the hazard to the lobster. However, it belongs to the same group of arthropods as the blue crab, has essentially the same physiology, and would presumably suffer the same effects. This would be true also of the stone crab and other crustaceans which have direct economic importance as human food.

The inshore waters – the bays, the sounds, the river estuaries, the tidal marshes – form an ecological unit of the utmost importance. They are linked so intimately and indispensably with the lives of many fishes, molluscs, and crustaceans that were they no longer habitable these sea foods would disappear from our tables.

Even among fishes that range widely in coastal waters, many depend upon protected inshore areas to serve as nursery and feeding grounds for their young. Baby tarpon are abundant in all that labyrinth of mangrove-lined streams and canals bordering the lower third of the western coast of Florida. On the Atlantic Coast the sea trout, croaker, spot, and drum spawn on sandy shoals off the inlets between the islands or 'banks' that lie like a protective chain off much of the coast south of New York. The young fish hatch and are carried through the inlets by the tides. In the bays and sounds – Currituck, Pamlico, Bogue, and many others – they find abundant food and grow rapidly. Without these nursery areas of warm, protected, food-rich waters the populations of these and many other species could not be maintained. Yet we are allowing pesticides to enter them via the rivers and by direct spraying over bordering marshlands. And the early stages of these fishes, even more than the adults, are especially susceptible to direct chemical poisoning.

Shrimp, too, depend on inshore feeding grounds for their young. One abundant and widely ranging species supports the entire commercial fishery of the southern Atlantic and Gulf states. Although spawning occurs at sea, the young come into the estuaries and bays when a few weeks old to undergo successive moults and changes of form. There they remain from May or June until autumn, feeding on the bottom detritus. In the entire period of their inshore life, the welfare of the shrimp populations and of the industry they support depends upon favourable conditions in the estuaries.

Do pesticides represent a threat to the shrimp fisheries and to the supply for the markets? The answer may be contained in recent laboratory experiments carried out by the Bureau of Commercial Fisheries. The insecticide tolerance of young commercial shrimp just past larval life was found to be exceedingly low – measured in parts per *billion* instead of the more commonly used standard of parts per million. For example, half the shrimp in one experiment were killed by dieldrin at a concentration of only 15 parts per billion. Other chemicals were even more toxic. Endrin, always one of the most deadly of the pesticides, killed half the shrimp at a concentration of only *half of one part per billion*.

The threat to oysters and clams is multiple. Again, the young stages are most vulnerable. These shellfish inhabit the bottoms of bays and sounds and tidal rivers from New England to Texas and sheltered areas of the Pacific Coast. Although sedentary in adult life, they discharge their spawn into the sea, where the young are free-living for a period of several weeks. On a summer day a fine-meshed tow net drawn behind a boat will collect, along with the other drifting plant and animal life that make up the plankton, the infinitely small, fragile-as-glass larvae of oysters and clams. No larger than grains of dust, these transparent larvae swim about in the surface waters, feeding on the microscopic plant life of the plankton. If the crop of minute sea vegetation fails, the young shellfish will starve. Yet pesticides may well destroy substantial quantities of plankton. Some of the herbicides in common use on lawns, cultivated fields, and roadsides and even in coastal marshes are extraordinarily toxic to the plant plankton which the larval molluscs use as food – some at only a few parts per billion.

The delicate larvae themselves are killed by very small quantities of many of the common insecticides. Even exposures to less than lethal quantities may in the end cause death of the larvae, for inevitably the growth rate is retarded. This prolongs the period the larvae must spend in the hazardous world of the plankton and so decreases the chance they will live to adulthood.

For adult molluscs there is apparently less danger of direct poisoning, at least by some of the pesticides. This is not necessarily reassuring, however. Oysters and clams may concentrate these poisons in their digestive organs and other tissues. Both types of shellfish are normally eaten whole and sometimes raw. Dr Philip Butler of the Bureau of Commercial Fisheries has pointed out an ominous parallel in that we may find ourselves in the same situation as the robins. The robins, he reminds us, did not die as a direct result of the spraying of DDT. They died because they had eaten earthworms that had already concentrated the pesticides in their tissues.

Although the sudden death of thousands of fish or crustaceans in some stream or pond as the direct and visible effect of insect control is dramatic and alarming, these unseen and as yet largely unknown and unmeasurable effects of pesticides reaching estuaries indirectly in streams and rivers may in the end be more disastrous. The whole situation is beset with questions for which there are at present no satisfactory answers. We know that pesticides contained in run-off from farms and forests are now being carried to the sea in the waters of many and perhaps all of the major rivers. But we do not know the identity of all the chemicals or their total quantity, and we do not at present have any dependable tests for identifying them in highly diluted state once they have reached the sea. Although we know that the chemicals have almost certainly undergone change during the long period of transit, we do not know whether the altered chemical is more toxic than the original or less. Another almost unexplored area is the question of interactions between chemicals, a question that becomes especially urgent when they enter the marine environment where so many different minerals are subjected to mixing and transport. All of these questions urgently require the precise answers that only extensive research can provide, yet funds for such purposes are pitifully small.

The fisheries of fresh and salt water are a resource of great importance, involving the interests and the welfare of a very large number of people. That they are now seriously threatened by the chemicals entering our waters can no longer be doubted. If we would divert to constructive research even a small fraction of the money spent each year on the development of ever more toxic sprays, we could find ways to use less dangerous materials and to keep poisons out of our waterways. When will the public become sufficiently aware of the facts to demand such action?

Chapter 10

Indiscriminately from the Skies

From small beginnings over farmlands and forests the scope of aerial spraying has widened and its volume has increased so that it has become what a British ecologist recently called 'an amazing rain of death' upon the surface of the earth. Our attitude towards poisons has undergone a subtle change. Once they were kept in containers marked with skull and crossbones; the infrequent occasions of their use were marked with utmost care that they should come in contact with the target and with nothing else. With the development of the new organic insecticides and the abundance of surplus planes after the Second World War, all this was forgotten. Although today's poisons are more dangerous than any known before, they have amazingly become something to be showered down indiscriminately from the skies. Not only the target insect or plant, but anything – human or non-human – within range of the chemical fallout may know the sinister touch of the poison. Not only forests and cultivated fields are sprayed, but towns and cities as well.

A good many people now have misgivings about the aerial distribution of lethal chemicals over millions of acres, and two mass-spraying campaigns undertaken in the late 1950s have done much to increase these doubts. These were the campaigns against the gypsy moth in the north-eastern states and the fire ant in the South. Neither is a native insect but both have been in this country for many years without creating a situation calling for desperate measures. Yet drastic action was suddenly taken against them, under the end-justifies-the-means philosophy that has too long directed the control divisions of our Department of Agriculture.

The gypsy moth programme shows what a vast amount of damage can be done when reckless large-scale treatment is substituted for local and moderate control. The campaign against the fire ant is a prime example of a campaign based on gross exaggeration of the need for control, blunderingly launched without scientific knowledge of the dosage of poison required to destroy the target or of its effects on other life. Neither programme has achieved its goal.

The gypsy moth, a native of Europe, has been in the United States for nearly a hundred years. In 1869 a French scientist, Leopold Trouvelot, accidentally allowed a few of these moths to escape from his laboratory in Medford, Massachusetts, where he was attempting to cross them with silkworms. Little by little the gypsy moth has spread throughout New England. The primary agent of its progressive spread is the wind; the larval, or caterpillar, stage is extremely light and can be carried to considerable heights and over great distances. Another means is the shipment of plants carrying the egg masses, the form in which the species exists over winter. The gypsy moth, which in its larval stage attacks the foliage of oak trees and a few other hardwoods for a few weeks each spring, now occurs in all the New England states. It also occurs sporadically in New Jersey, where it was introduced in 1911 on a shipment of spruce trees from Holland, and in Michigan, where its method of entry is not known. The New England hurricane of 1938 carried it into Pennsylvania and New York, but the Adirondacks have generally served as a barrier to its westward advance, being forested with species not attractive to it.

The task of confining the gypsy moth to the north-eastern corner of the country has been accomplished by a variety of methods, and in the nearly one hundred years since its arrival on this continent the fear that it would invade the great hardwood forests of the southern Appalachians has not been justified. Thirteen parasites and predators were imported from abroad and successfully established in New England. The Agriculture Department itself has credited these importations with appreciably reducing the frequency and destructiveness of gypsy moth outbreaks. This natural control, plus quarantine measures and local spraying, achieved what the Department in 1955 described as 'outstanding restriction of distribution and damage'.

Yet only a year after expressing a satisfaction with the state of affairs, its Plant Pest Control Division embarked on a programme calling for the blanket spraying of several million acres a year with the announced intention of eventually 'eradicating' the gypsy moth. ('Eradication' means the complete and final extinction or extermination of a species throughout its range. Yet as successive programmes have failed, the Department has found it necessary to speak of second or third 'eradications' of the same species in the same area.)

The Department's all-out chemical war on the gypsy moth began on an ambitious scale. In 1956 nearly a million acres were sprayed in the states of Pennsylvania, New Jersey, Michigan, and New York. Many complaints of damage were made by people in the sprayed areas. Conservationists became increasingly disturbed as the pattern of spraying huge areas began to establish itself. When plans were announced for spraying three million acres in 1957 opposition became even stronger. State and federal agriculture officials characteristically shrugged off individual complaints as unimportant.

The Long Island area included within the gypsy moth spraying in 1957 consisted chiefly of heavily populated towns and suburbs and of some coastal areas with bordering salt marsh. Nassau County, Long Island, is the most densely settled county in New York apart from New York City itself. In what seems the height of absurdity, the 'threat of infestation of the New York City metropolitan area' has been cited as an important justification of the programme. The gypsy moth is a forest insect, certainly not an inhabitant of cities. Nor does it live in meadows, cultivated fields, gardens, or marshes. Nevertheless, the planes hired by the United States Department of Agriculture and the New York Department of Agriculture and Markets in 1957 showered down the prescribed DDT-in-fuel-oil with impartiality. They sprayed truck gardens and dairy farms, fish ponds and salt marshes. They sprayed the quarter-acre plots of suburbia, drenching a housewife making a desperate effort to cover her garden before the roaring plane reached her, and showering insecticide over children at play and commuters at railway stations. At Setauket a fine quarter horse drank from a trough in a field which the planes had sprayed; ten hours later it was dead. Motor-cars were spotted with the oily mixture;

flowers and shrubs were ruined. Birds, fish, crabs, and useful insects were killed.

A group of Long Island citizens led by the world-famous ornithologist Robert Cushman Murphy had sought a court injunction to prevent the 1957 spraying. Denied a preliminary injunction, the protesting citizens had to suffer the prescribed drenching with DDT, but thereafter persisted in efforts to obtain a permanent injunction. But because the act had already been performed the courts held that the petition for an injunction was 'moot'. The case was carried all the way to the Supreme Court, which declined to hear it. Justice William O. Douglas, strongly dissenting from the decision not to review the case, held that 'the alarms that many experts and responsible officials have raised about the perils of DDT underline the public importance of this case'.

The suit brought by the Long Island citizens at least served to focus public attention on the growing trend to mass application of insecticides, and on the power and inclination of the control agencies to disregard supposedly inviolate property rights of private citizens.

The contamination of milk and of farm produce in the course of the gypsy moth spraying came as an unpleasant surprise to many people. What happened on the 200-acre Waller farm in northern Westchester County, New York, was revealing. Mrs Waller had specifically requested Agriculture officials not to spray her property; because it would be impossible to avoid the pastures in spraying the woodlands. She offered to have the land checked for gypsy moths and to have any infestation destroyed by spot spraying. Although she was assured that no farms would be sprayed, her property received two direct sprayings and, in addition, was twice subjected to drifting spray. Milk samples taken from the Wallers' pure-bred Guernsey cows forty-eight hours later contained DDT in the amount of 14 parts per million. Forage samples from fields where the cows had grazed were, of course, contaminated also. Although the county Health Department was notified, no instructions were given that the milk should not be marketed. This situation is unfortunately typical of the lack of consumer protection that is all too common. Although the Food and Drug Administration permits no residues of pesticides in milk, its restrictions are not only inadequately policed but they apply solely to inter-state shipments. State and county officials are under no compulsion to follow the federal pesticides tolerances unless local laws happen to conform – and they seldom do.

Truck gardeners also suffered. Some leaf crops were so burned and spotted as to be unmarketable. Others carried heavy residues; a sample of peas analysed at Cornell University's Agricultural Experiment Station contained 14 to 20 parts per million of DDT. The legal maximum is 7 parts per million. Growers therefore had to sustain heavy losses or find themselves in the position of selling produce carrying illegal residues. Some of them sought and collected damages.

As the aerial spraying of DDT increased, so did the number of suits filed in the courts. Among them were suits brought by beekeepers in several areas of New York State. Even before the 1957 spraying, the beekeepers had suffered heavily from use of DDT in orchards. 'Up to 1953 I had regarded as gospel everything that emanated from the U.S. Department of Agriculture and the agricultural colleges,' one of them remarked bitterly. But in May of that year this man lost 800 colonies after the state had sprayed a large area. So widespread and heavy was the loss that fourteen other beekeepers joined him in suing the state for a quarter of a million dollars in damages. Another beekeeper, whose 400 colonies were incidental targets of the 1957 spray, reported that 100 per cent of the field force of bees (the workers out gathering nectar and pollen for the hives) had been killed in forested areas and up to 50 per cent in farming areas sprayed less intensively. 'It is a very distressing thing,' he wrote, 'to walk into a yard in May and not hear a bee buzz.'

The gypsy moth programmes were marked by many acts of irresponsibility. Because the spray planes were paid by the gallon rather than by the acre there was no effort to be conservative, and many properties were sprayed not once but several times. Contracts for aerial spraying were in at least one case awarded to an out-of-state firm with no local address, which had not complied with the legal requirement of registering with state officials for the purpose of establishing legal responsibility. In this exceedingly slippery situation, citizens who suffered direct financial loss from damage to apple orchards or bees discovered that there was no one to sue.

After the disastrous 1957 spraying the programme was abruptly and drastically curtailed, with vague statements about 'evaluating' previous work and testing alternative insecticides. Instead of the 3½ million acres sprayed in 1957, the treated areas fell to ½ million in 1958 and to about 100,000 acres in 1959, 1960, and 1961. During this interval, the control agencies must have found news from Long Island disquieting. The gypsy moth had reappeared there in numbers. The expensive spraying operation that had cost the Department dearly in public confidence and good will – the operation that was intended to wipe out the gypsy moth for ever – had in reality accomplished nothing at all.

Meanwhile, the Department's Plant Pest Control men had temporarily forgotten gypsy moths, for they had been busy launching an even more ambitious programme in the South. The word 'eradication' still came easily from the Department's mimeograph machines; this time the press releases were promising the eradication of the fire ant.

The fire ant, an insect named after its fiery sting, seems to have entered the United States from South America by way of the port of Mobile, Alabama, where it was discovered shortly after the end of the First World War. By 1928 it had spread into the suburbs of Mobile and thereafter continued an invasion that has now carried it into most of the southern states.

During most of the forty-odd years since its arrival in the United States the fire ant seems to have attracted little attention. The states where it was most abundant considered it a nuisance, chiefly because it builds large nests or mounds a foot or more high. These may hamper the operation of farm machinery. But only two states listed it among their twenty most important insect pests, and these placed it near the bottom of the list. No official or private concern seems to have been felt about the fire ant as a menace to crops or livestock.

With the development of chemicals of broad lethal powers, there came a sudden change in the official attitude towards the fire ant. In 1957 the United States Department of Agriculture launched one of the most remarkable publicity campaigns in its history. The fire ant suddenly became the target of a barrage of government releases, motion pictures, and government-inspired stories portraying it as a despoiler of southern agriculture and a killer of birds, livestock, and man. A mighty campaign was announced, in which the federal government in cooperation with the afflicted states would ultimately treat some 20,000,000 acres in nine southern states.

United States pesticide makers appear to have tapped a sales bonanza in the increasing numbers of broad-scale pest elimination programmes conducted by the U.S. Department of Agriculture,

cheerfully reported one trade journal in 1958, as the fire ant programme got under way.

Never has any pesticide programme been so thoroughly and deservedly damned by practically everyone except the beneficiaries of this 'sales bonanza'. It is an outstanding example of an ill-conceived, badly executed, and thoroughly detrimental experiment in the mass control of insects, an experiment so expensive in dollars, in destruction of animal life, and in loss of public confidence in the Agriculture Department that it is incomprehensible that any funds should still be devoted to it.

Congressional support of the project was initially won by representations that were later discredited. The fire ant was pictured as a serious threat to southern agriculture through destruction of crops and to wildlife because of attacks on the young of ground-nesting birds. Its sting was said to make it a serious menace to human health.

Just how sound were these claims? The statements made by Department witnesses seeking appropriations were not in accord with those contained in key publications of the Agriculture Department. The 1957 bulletin *Insecticide Recommendations... for the Control of Insects Attacking Crops and Livestock* did not so much as mention the fire ant – an extraordinary omission if the Department believes its own propaganda. Moreover, its encyclopaedic *Yearbook* for 1952, which was devoted to insects, contained only one short paragraph on the fire ant out of its half-million words of text.

Against the Department's undocumented claim that the fire

ant destroys crops and attacks livestock is the careful study of the Agricultural Experiment Station in the state that has had the most intimate experience with this insect, Alabama. According to Alabama scientists, 'damage to plants in general is rare'. Dr F. S. Arant, an entomologist at the Alabama Polytechnic Institute and in 1961 president of the Entomological Society of America, states that his department 'has not received a single report of damage to plants by ants in the past five years.... No damage to livestock has been observed.' These men, who have actually observed the ants in the field and in the laboratory, say that the fire ants feed chiefly on a variety of other insects, many of them considered harmful to man's interests. Fire ants have been observed picking larvae of the boll weevil off cotton. Their mound-building activities serve a useful purpose in aerating and draining the soil. The Alabama studies have been substantiated by investigations at the Mississippi State University, and are far more impressive than the Agriculture Department's evidence, apparently based either on conversations with farmers, who may easily mistake one ant for another, or on old research. Some entomologists believe that the ant's food habits have changed as it has become more abundant, so that observations made several decades ago have little value now.

The claim that the ant is a menace to health and life also bears considerable modification. The Agriculture Department sponsored a propaganda film (to gain support for its programme) in which horror scenes were built around the fire ant's sting. Admittedly this is painful and one is well advised to avoid being stung, just as one ordinarily avoids the sting of wasp or bee. Severe reactions may occasionally occur in sensitive individuals, and medical literature records one death possibly, though not definitely, attributable to fire ant venom. In contrast to this, the Office of Vital Statistics records thirty-three deaths in 1959 alone from the sting of bees and wasps. Yet no one seems to have proposed 'eradicating' these insects. Again, local evidence is most convincing. Although the fire ant has inhabited Alabama for forty years and is most heavily concentrated there, the Alabama State Health Officer declares that 'there has never been recorded in Alabama a human death resulting from the bites of imported fire ants', and considers the medical cases resulting from the bites of fire ants 'incidental'. Ant mounds on lawns or playgrounds may create a situation where children are likely to be stung, but this is hardly an excuse for drenching millions of acres with poisons. These situations can easily be handled by individual treatment of the mounds.

Damage to game birds was also alleged, without supporting evidence. Certainly a man well qualified to speak on this issue is the leader of the Wildlife Research Unit at Auburn, Alabama, Dr Maurice F. Baker, who has had many years' experience in the area. But Dr Baker's opinion is directly opposite to the claims of the Agriculture Department. He declares:

In south Alabama and north-west Florida we are able to have excellent hunting and bobwhite populations coexistent with heavy populations of the imported fire ant ... in the almost forty years that south Alabama has had the fire ant, game populations have shown a steady and very substantial increase. Certainly, if the imported fire ant were a serious menace to wildlife, these conditions could not exist.

What would happen to wildlife as a result of the insecticide used against the ants was another matter. The chemicals to be used were dieldrin and heptachlor, both relatively new. There was little experience of field use for either, and no one knew what their effects would be on wild birds, fishes, or mammals when applied on a massive scale. It was known, however, that both poisons were many times more toxic than DDT, which had been used by that time for approximately a decade, and had killed some birds and many fish even at a rate of 1 pound per acre. And the dosage of dieldrin and heptachlor was heavier – 2 pounds to the acre under most conditions, or 3 pounds of dieldrin if the white-fringed beetle was also to be controlled. In terms of their effects on birds, the prescribed use of heptachlor would be equivalent to 20 pounds of DDT to the acre, that of dieldrin 120 pounds!

Urgent protests were made by most of the state conservation departments, by national conservation agencies, and by ecologists and even by some entomologists, calling upon the then Secretary of Agriculture, Ezra Benson, to delay the programme at least until some research had been done to determine the effects of heptachlor and dieldrin on wild and domestic animals and to find the minimum amount that would control the ants.

The protests were ignored and the programme was launched in 1958. A million acres were treated the first year. It was clear that any research would be in the nature of a post mortem.

As the programme continued, facts began to accumulate from studies made by biologists of state and federal wildlife agencies and several universities. The studies revealed losses running all the way up to complete destruction of wildlife on some of the treated areas. Poultry, livestock, and pets were also killed. The Agriculture Department brushed away all evidence of damage as exaggerated and misleading.

The facts, however, continue to accumulate. In Hardin County, Texas, for example, opossums, armadillos, and an abundant raccoon population virtually disappeared after the chemical was laid down. Even the second autumn after treatment these animals were scarce. The few raccoons then found in the area carried residues of the chemical in their tissues.

Dead birds found in the treated areas had absorbed or swallowed the poisons used against the fire ants, a fact clearly shown by chemical analysis of their tissues. (The only bird surviving in any numbers was the house sparrow, which in other areas too has given some evidence that it may be relatively immune.) On a tract in Alabama treated in 1959 half of the birds were killed. Species that live on the ground or frequent low vegetation suffered 100 per cent mortality. Even a year after treatment, a spring die-off of songbirds occurred and much good nesting territory lay silent and unoccupied. In Texas, dead blackbirds, dickcissels, and meadowlarks were found at the nests, and many nests were deserted. When specimens of dead birds from Texas, Louisiana, Alabama, Georgia, and Florida were sent to the Fish and Wildlife Service for analysis, more than 90 per cent were found to contain residues of dieldrin or a form of heptachlor, in amounts up to 38 parts per million.

Woodcocks, which winter in Louisiana but breed in the north, now carry the taint of the fire ant poisons in their bodies. The source of this contamination is clear. Woodcocks feed heavily on earthworms, which they probe for with their long bills. Surviving worms in Louisiana were found to have as much as 20 parts per million of heptachlor in their tissues six to ten months after treatment of the area. A year later they had up to

10 parts per million. The consequences of the sub-lethal poisoning of the woodcock are now seen in a marked decline in the proportion of young birds to adults, first observed in the season after fire ant treatments began.

Some of the most upsetting news for southern sportsmen concerned the bobwhite quail. This bird, a ground nester and forager, was all but eliminated on treated areas. In Alabama, for example, biologists of the Alabama Co-operative Wildlife Research Unit conducted a preliminary census of the quail population in a 3,600-acre area that was scheduled for treatment. Thirteen resident coveys – 121 quail – ranged over the area. Two weeks after treatment only dead quail could be found. All specimens sent to the Fish and Wildlife Service for analysis were found to contain insecticides in amounts sufficient to cause their death. The Alabama findings were duplicated in Texas, where a 2,500-acre area treated with heptachlor lost all of its quail. Along with the quail went 90 per cent of the songbirds. Again, analysis revealed the presence of heptachlor in the tissues of dead birds.

In addition to quail, wild turkeys were seriously reduced by the fire ant programme. Although eighty turkeys had been counted on an area in Wilcox County, Alabama, before heptachlor was applied, none could be found the summer after treatment – none, that is, except a clutch of unhatched eggs and one dead poult. The wild turkeys may have suffered the same fate as their domestic brethren, for turkeys on farms in the area treated with chemicals also produced few young. Few eggs hatched and almost no young survived. This did not happen on nearby untreated areas.

The fate of the turkeys was by no means unique. One of the most widely known and respected wildlife biologists in the country, Dr Clarence Cottam, called on some of the farmers whose property had been treated. Besides remarking that 'all the little tree birds' seemed to have disappeared after the land had been treated, most of these people reported losses of livestock, poultry, and household pets. One man was 'irate against the control workers', Dr Cottam reported,

as he said he buried or otherwise disposed of nineteen carcasses of his cows that had been killed by the poison and he knew of three or four additional cows that died as a result of the same treatment. Calves died that had been given only milk since birth.

The people Dr Cottam interviewed were puzzled by what had happened in the months following the treatment of their land. One woman told him she had set several hens after the surrounding land had been covered with poison, 'and for reasons she did not understand very few young were hatched or survived'. Another farmer

raises hogs and for fully nine months after the broadcast of poisons, he could raise no young pigs. The litters were born dead or they died after birth.

A similar report came from another, who said that out of thirty-seven litters that might have numbered as many as 250 young, only thirty-one little pigs survived. This man had also been quite unable to raise chickens since the land was poisoned.

The Department of Agriculture has consistently denied livestock losses related to the fire ant programme. However, a veterinarian in Bainbridge, Georgia, Dr Otis L. Poitevint, who was called upon to treat many of the affected animals, has summarized his reasons for attributing the deaths to the insecticide as follows. Within a period of two weeks to several months after the fire ant poison was applied, cattle, goats, horses, chickens, and birds and other wildlife began to suffer an often fatal disease of the nervous system. It affected only animals that had access to contaminated food or water. Stabled animals were not affected. The condition was seen only in areas treated for fire ants. Laboratory tests for disease were negative. The symptoms observed by Dr Poitevint and other veterinarians were those described in authoritative texts as indicating poisoning by dieldrin or heptachlor.

Dr Poitevint also described an interesting case of a two-month-old calf that showed symptoms of poisoning by heptachlor. The animal was subjected to exhaustive laboratory tests. The only significant finding was the discovery of 79 parts per million of heptachlor in its fat. But it was five months since the poison had been applied. Did the calf get it directly from grazing or indirectly from its mother's milk or even before birth?

If from the milk [asked Dr Poitevint], why were not special precautions taken to protect our children who drank milk from local dairies?

Dr Poitevint's report brings up a significant problem about the contamination of milk. The area included in the fire ant programme is predominantly fields and croplands. What about the dairy cattle that graze on these lands? In treated fields the grasses will inevitably carry residues of heptachlor in one of its forms, and if the residues are eaten by the cows the poison will appear in the milk. This direct transmission into milk had been demonstrated experimentally for heptachlor in 1955, long before the control programme was undertaken, and was later reported for dieldrin, also used in the fire ant programme.

The Department of Agriculture's annual publications now list heptachlor and dieldrin among the chemicals that make forage plants unsuitable for feeding to dairy animals or animals being finished for slaughter, yet the control divisions of the Department promote programmes that spread heptachlor and dieldrin over substantial areas of grazing land in the south. Who is safeguarding the consumer to see that no residues of dieldrin or heptachlor are appearing in milk? The United States Department of Agriculture would doubtless answer that it has advised farmers to keep milk cows out of treated pastures for thirty to ninety days. Given the small size of many of the farms and the large-scale nature of the programme – much of the chemical applied by planes – it is extremely doubtful that this recommendation was followed or could be. Nor is the prescribed period adequate in view of the persistent nature of the residues.

The Food and Drug Administration, although frowning on the presence of any pesticide residues in milk, has little authority in this situation. In most of the states included in the fire ant programme the dairy industry is small and its products do not cross state lines. Protection of the milk supply endangered by a federal programme is therefore left to the states themselves. Inquiries addressed to the health officers or other appropriate officials of Alabama, Louisiana, and Texas in 1959 revealed that no tests had been made and that it simply was not known whether the milk was contaminated with pesticides or not.

Meanwhile, after rather than before the control programme was launched, some research into the peculiar nature of heptachlor was done. Perhaps it would be more accurate to say that someone looked up the research already published, since the basic fact that brought about belated action by the federal government had been discovered several years before, and should have influenced the initial handling of the programme. This is the fact that heptachlor, after a short period in the tissues of animals or plants or in the soil, assumes a considerably more toxic form known as heptachlor epoxide. The epoxide is popularly described as 'an oxidation product' produced by weathering. The fact that this transformation could occur had been known since 1952, when the Food and Drug Administration discovered that female rats, fed 30 parts per million of heptachlor, had stored 165 parts per million of the more poisonous epoxide only two weeks later.

These facts were allowed to come out of the obscurity of biological literature in 1959, when the Food and Drug Administration took action which had the effect of banning any residues of heptachlor or its epoxide on food. This ruling put at least a temporary damper on the programme; although the Agriculture Department continued to press for its annual appropriations for fire ant control, local agricultural agents became increasingly reluctant to advise farmers to use chemicals which would probably result in their crops being legally unmarketable.

In short, the Department of Agriculture embarked on its programme without even elementary investigation of what was already known about the chemical to be used – or if it investigated, it ignored the findings. It must also have failed to do preliminary research to discover the minimum amount of the chemical that would accomplish its purpose. After three years of heavy dosages, it abruptly reduced the rate of application of heptachlor from 2 pounds to $1\frac{1}{4}$ pounds per acre in 1959, later on to $\frac{1}{2}$ pound per acre, applied in two treatments of $\frac{1}{4}$ pound each, three to six months apart. An official of the Department explained that 'an aggressive methods improvement programme' showed the lower rate to be effective. Had this information been acquired before the programme was launched, a vast amount of damage could have been avoided and the taxpayers could have been saved a great deal of money.

In 1959, perhaps in an attempt to offset the growing dissatis-faction with the programme, the Agriculture Department offered the chemicals free to Texas landowners who would sign a release absolving federal, state, and local governments of responsibility for damage. In the same year the State of Alabama, alarmed and angry at the damage done by the chemicals, refused to appropri-ate any further funds for the project. One of its officials character-ized the whole programme as

ill advised, hastily conceived, poorly planned, and a glaring example of riding roughshod over the responsibilities of other public and private agencies.

Despite the lack of state funds, federal money continued to trickle into Alabama, and in 1961 the legislature was again per-suaded to make a small appropriation. Meanwhile, farmers in Louisiana showed growing reluctance to sign up for the project as it became evident that use of chemicals against the fire ant was causing an upsurge of insects destructive to sugar-cane. Moreover, the programme was obviously accomplishing nothing. Its dismal state was tersely summarized in the spring of 1962 by the director of entomology research at Louisiana State Univer-sity Agricultural Experiment Station, Dr L. D. Newsom:

The imported fire ant 'eradication' programme which has been conducted by state and federal agencies is thus far a failure. There are more infested acres in Louisiana now than when the programme began.

A swing to more sane and conservative methods seems to have begun. Florida, reporting that 'there are more fire ants in Florida now than there were when the programme started', announced it was abandoning any idea of a broad eradication programme and would instead concentrate on local control.

Effective and inexpensive methods of local control have been known for years. The mound-building habit of the fire ant makes the chemical treatment of individual mounds a simple matter. Cost of such treatment is about one dollar per acre. For situa-tions where mounds are numerous and mechanized methods are desirable, a cultivator which first levels and then applies chemical directly to the mounds has been developed by Mississippi's

Agricultural Experiment Station. The method gives 90 to 95 per cent control of the ants. Its cost is only $·23 per acre. The Agriculture Department's mass control programme, on the other hand, cost about $3.50 per acre – the most expensive, the most damaging, and the least effective programme of all.

Chapter 11

Beyond the Dreams of
the Borgias

The contamination of our world is not alone a matter of mass spraying. Indeed, for most of us this is of less importance than the innumerable small-scale exposures to which we are subjected day by day, year after year. Like the constant dripping of water that in turn wears away the hardest stone, this birth-to-death contact with dangerous chemicals may in the end prove disastrous. Each of these recurrent exposures, no matter how slight, contributes to the progressive build-up of chemicals in our bodies and so to cumulative poisoning. Probably no person is immune to contact with this spreading contamination unless he lives in the most isolated situation imaginable. Lulled by the soft sell and the hidden persuader, the average citizen is seldom aware of the deadly materials with which he is surrounding himself; indeed, he may not realize he is using them at all.

So thoroughly has the age of poisons become established that anyone may walk into a store and, without questions being asked, buy substances of far greater death-dealing power than the medicinal drug for which he may be required to sign a 'poison book' in the pharmacy next door. A few minutes' research in any supermarket is enough to alarm the most stout-hearted customer – provided, that is, he has even a rudimentary knowledge of the chemicals presented for his choice.

If a huge skull and crossbones were suspended above the insecticide department the customer might at least enter it with the respect normally accorded death-dealing materials. But instead the display is homey and cheerful, and, with the pickles and olives across the aisle and the bath and laundry soaps adjoining, the rows upon rows of insecticides are displayed. Within easy reach of a child's exploring hand are chemicals in *glass* containers. If dropped to the floor by a child or careless adult everyone near by could be splashed with the same chemical that has sent spraymen using it into convulsions. These hazards, of course, follow the purchaser right into his home. A can of a moth-proofing material containing DDD, for example, carries in very fine print the warning that its contents are under pressure and that it may burst if exposed to heat or open flame. A common insecticide for household use, including assorted uses in the kitchen, is chlordane. Yet the Food and Drug Administration's chief pharmacologist has declared the hazard of living in a house sprayed with chlordane to be 'very great'. Other household preparations contain the even more toxic dieldrin.

Use of poisons in the kitchen is made both attractive and easy. Kitchen shelf paper, white or tinted to match one's colour scheme, may be impregnated with insecticide, not merely on one but on both sides. Manufacturers offer us do-it-yourself booklets on how to kill bugs. With push-button ease, one may send a fog of dieldrin into the most inaccessible nooks and crannies of cabinets, corners, and baseboards.

If we are troubled by mosquitoes, chiggers, or other insect pests on our persons we have a choice of innumerable lotions, creams, and sprays for application to clothing or skin. Although we are warned that some of these will dissolve varnish, paint, and synthetic fabrics, we are presumably to infer that the human skin is impervious to chemicals. To make certain that we shall at all times be prepared to repel insects, an exclusive New York store advertises a pocket-sized insecticide dispenser, suitable for the purse or for beach, golf, or fishing gear.

We can polish our floors with a wax guaranteed to kill any insect that walks over it. We can hang strips impregnated with the chemical lindane in our closets and garment bags or place them in our bureau drawers for a half-year's freedom from worry over moth damage. The advertisements contain no suggestion that lindane is dangerous. Neither do the ads for an electronic device that dispenses lindane fumes – we are told that it is safe and odourless. Yet the truth of the matter is that the American Medical Association considers lindane vaporizers so dangerous that it conducted an extended campaign against them in its *Journal*.

The Department of Agriculture, in a *Home and Garden*

Bulletin, advises us to spray our clothing with oil solutions of DDT, dieldrin, chlordane, or any of several other moth-killers. If excessive spraying results in a white deposit of insecticide on the fabric, this may be removed by brushing, the Department says, omitting to caution us to be careful where and how the brushing is done. All these matters attended to, we may round out our day with insecticides by going to sleep under a moth-proof blanket impregnated with dieldrin.

Gardening is now firmly linked with the super-poisons. Every hardware store, garden-supply shop, and supermarket has rows of insecticides for every conceivable horticultural situation. Those who fail to make wide use of this array of lethal sprays and dusts are by implication remiss, for almost every newspaper's garden page and the majority of the gardening magazines take their use for granted.

So extensively are even the rapidly lethal organic phosphorus insecticides applied to lawns and ornamental plants that in 1960 the Florida State Board of Health found it necessary to forbid the commercial use of pesticides in residential areas by anyone who had not first obtained a permit and met certain requirements. A number of deaths from parathion had occurred in Florida before this regulation was adopted.

Little is done, however, to warn the gardener or home-owner that he is handling extremely dangerous materials. On the contrary, a constant stream of new gadgets make it easier to use poisons on lawn and garden – and increase the gardener's contact with them. One may get a jar-type attachment for the garden hose, for example, by which such extremely dangerous chemicals as chlordane or dieldrin are applied as one waters the lawn. Such a device is not only a hazard to the person using the hose; it is also a public menace. The *New York Times* found it necessary to issue a warning on its garden page to the effect that unless special protective devices were installed poisons might get into the water supply by back siphonage. Considering the number of such devices that are in use, and the scarcity of warnings such as this, do we need to wonder why our public waters are contaminated?

As an example of what may happen to the gardener himself, we might look at the case of a physician – an enthusiastic sparetime gardener – who began using DDT and then malathion on his shrubs and lawn, making regular weekly applications. Sometimes he applied the chemicals with a hand spray, sometimes with an attachment to his hose. In doing so, his skin and clothing were often soaked with spray. After about a year of this sort of thing, he suddenly collapsed and was hospitalized. Examination of a biopsy specimen of fat showed an accumulation of 23 parts per million of DDT. There was extensive nerve damage, which his physicians regarded as permanent. As time went on he lost weight, suffered extreme fatigue, and experienced a peculiar muscular weakness, a characteristic effect of malathion. All of these persisting effects were severe enough to make it difficult for the physician to carry on his practice.

Besides the once-innocuous garden hose, power mowers also have been fitted with devices for the dissemination of pesticides, attachments that will dispense a cloud of vapour as the home-owner goes about the task of mowing his lawn. So to the potentially dangerous fumes from petrol are added the finely divided particles of whatever insecticide the probably unsuspecting suburbanite has chosen to distribute, raising the level of air pollution above his own grounds to something few cities could equal.

Yet little is said about the hazards of the fad of gardening by poisons, or of insecticides used in the home; warnings on labels are printed so inconspicuously in small type that few take the trouble to read or follow them. An industrial firm recently undertook to find out just *how* few. Its survey indicated that fewer than fifteen people out of a hundred of those using insecticide aerosols and sprays are even aware of the warnings on the containers.

The mores of suburbia now dictate that crabgrass must go at whatever cost. Sacks containing chemicals designed to rid the lawn of such despised vegetation have become almost a status symbol. These weed-killing chemicals are sold under brand names that never suggest their identity or nature. To learn that they contain chlordane or dieldrin one must read exceedingly fine print placed on the least conspicuous part of the sack. The descriptive literature that may be picked up in any hardware or garden-supply store seldom if ever reveals the true hazard involved in handling or applying the material. Instead, the typical illustration portrays a happy family scene, father and son smilingly preparing to apply the chemical to the lawn, small children tumbling over the grass with a dog.

The question of chemical residues on the food we eat is a hotly debated issue. The existence of such residues is either played down by the industry as unimportant or is flatly denied. Simultaneously, there is a strong tendency to brand as fanatics or cultists all who are so perverse as to demand that their food be free of insect poisons. In all this cloud of controversy, what are the actual facts?

It has been medically established that, as common sense would tell us, persons who lived and died before the dawn of the DDT era (about 1942) contained no trace of DDT or any similar material in their tissues. As mentioned in Chapter 3, samples of body fat collected from the general population between 1954 and 1956 averaged from 5·3 to 7·4 parts per million of DDT. There is some evidence that the average level has risen since then to a consistently higher figure, and individuals with occupational or other special exposures to insecticides of course store even more.

Among the general population with no known gross exposures to insecticides it may be assumed that much of the DDT stored in fat deposits has entered the body in food. To test this assumption, a scientific team from the United States Public Health Service sampled restaurant and institutional meals. *Every meal sampled contained DDT*. From this the investigators concluded, reasonably enough, that 'few if any foods can be relied upon to be entirely free of DDT'.

The quantities in such meals may be enormous. In a separate Public Health Service study, analysis of prison meals disclosed such items as stewed dried fruit containing 69·6 parts per million and bread containing 100·9 parts per million of DDT!

In the diet of the average home, meats and any products derived from animal fats contain the heaviest residues of chlorinated hydrocarbons. This is because these chemicals are soluble in fat. Residues on fruits and vegetables tend to be somewhat less. These are little affected by washing – the only remedy is to remove and discard all outside leaves of such vegetables as lettuce or cabbage, to peel fruit and to use no skins or outer covering whatever. Cooking does not destroy residues.

Milk is one of the few foods in which no pesticide residues are permitted by Food and Drug Administration regulations. In actual fact, however, residues turn up whenever a check is made. They are heaviest in butter and other manufactured dairy products. A check of 461 samples of such products in 1960 showed that a third contained residues, a situation which the Food and Drug Administration characterized as 'far from encouraging'.

To find a diet free from DDT and related chemicals, it seems one must go to a remote and primitive land, still lacking the amenities of civilization. Such a land appears to exist, at least marginally, on the far Arctic shores of Alaska – although even there one may see the approaching shadow. When scientists investigated the native diet of the Eskimos in this region it was found to be free from insecticides. The fresh and dried fish; the fat, oil, or meat from beaver, beluga, caribou, moose, oogruk, polar bear, and walrus; cranberries, salmonberries and wild rhubarb all had so far escaped contamination. There was only one exception – two white owls from Point Hope carried small amounts of DDT, perhaps acquired in the course of some migratory journey.

When some of the Eskimos themselves were checked by analysis of fat samples, small residues of DDT were found (0 to 1·9 parts per million). The reason for this was clear. The fat samples were taken from people who had left their native villages to enter the United States Public Health Service Hospital in Anchorage for surgery. There the ways of civilization prevailed, and the meals in this hospital were found to contain as much DDT as those in the most populous city. For their brief stay in civilization the Eskimos were rewarded with a taint of poison.

The fact that every meal we eat carries its load of chlorinated hydrocarbons is the inevitable consequence of the almost universal spraying or dusting of agricultural crops with these poisons. If the farmer scrupulously follows the instructions on the labels, his use of agricultural chemicals will produce no residues larger than are permitted by the Food and Drug Administration. Leaving aside for the moment the question whether these legal residues are as 'safe' as they are represented to be, there remains the well-known fact that farmers very frequently exceed the prescribed dosages, use the chemical too close to the time of harvest, use several insecticides where one would do, and in other ways display the common human failure to read the fine print.

Even the chemical industry recognizes the frequent misuse of insecticides and the need for education of farmers. One of its leading trade journals recently declared that

many users do not seem to understand that they may exceed insecticide tolerances if they use higher dosages than recommended. And haphazard use of insecticides on many crops may be based on farmers' whims.

The files of the Food and Drug Administration contain records of a disturbing number of such violations. A few examples will serve to illustrate the disregard of directions: a lettuce farmer who applied not one but eight different insecticides to his crop within a short time of harvest, a shipper who had used the deadly parathion on celery in an amount five times the recommended maximum, growers using endrin – most toxic of all the chlorinated hydrocarbons – on lettuce although no residue was allowable, spinach sprayed with DDT a week before harvest.

There are also cases of chance or accidental contamination. Large lots of green coffee in burlap bags have become contaminated while being transported by vessels also carrying a cargo of insecticides. Packaged foods in warehouses are subjected to repeated aerosol treatments with DDT, lindane, and other insecticides, which may penetrate the packaging materials and occur in measurable quantities on the contained foods. The longer the food remains in storage, the greater the danger of contamination.

To the question 'But doesn't the government protect us from such things?' the answer is, 'Only to a limited extent'. The activities of the Food and Drug Administration in the field of consumer protection against pesticides are severely limited by two facts. The first is that it has jurisdiction only over foods shipped in inter-state commerce; foods grown and marketed within a state are entirely outside its sphere of authority, no matter what the violation. The second and critically limiting fact is the small number of inspectors on its staff – fewer than 600 men for all its varied work. According to a Food and Drug official, only an infinitesimal part of the crop products moving in inter-state commerce – far less than 1 per cent – can be checked with existing facilities, and this is not enough to have statistical significance. As for food produced and sold within a state, the situation is even worse, for most states have woefully inadequate laws in this field.

The system by which the Food and Drug Administration establishes maximum permissible limits of contamination, called 'tolerances', has obvious defects. Under the conditions prevailing it provides mere paper security and promotes a completely unjustified impression that safe limits have been established and are being adhered to. As to the safety of allowing a sprinkling of poisons on our food – a little on this, a little on that – many people contend, with highly persuasive reasons, that no poison is safe or desirable on food. In setting a tolerance level the Food and Drug Administration reviews tests of the poison on laboratory animals and then establishes a maximum level of contamination that is much less than required to produce symptoms in the test animal. This system, which is supposed to ensure safety, ignores a number of important facts. A laboratory animal, living under controlled and highly artificial conditions, consuming a given amount of a specific chemical, is very different from a human being whose exposures to pesticides are not only multiple but for the most part unknown, unmeasurable and uncontrollable. Even if 7 parts per million of DDT on the lettuce in his luncheon salad were 'safe', the meal includes other foods, each with allowable residues, and the pesticides on his food are, as we have seen, only a part, and possibly a small part, of his total exposure. This piling up of chemicals from many different sources creates a total exposure that cannot be measured. It is meaningless, therefore, to talk about the 'safety' of any specific amount of residue.

And there are other defects. Tolerances have sometimes been established against the better judgement of Food and Drug Administration scientists, as in the case cited on pages 197 ff., or they have been established on the basis of inadequate knowledge of the chemical concerned. Better information has led to later reduction or withdrawal of the tolerance, but only after the public has been exposed to admittedly dangerous levels of the chemical for months or years. This happened when heptachlor was given a tolerance that later had to be revoked. For some chemicals no practical field method of analysis exists before a chemical is registered for use. Inspectors are therefore frustrated in their search for residues. This difficulty greatly hampered the work on the 'cranberry chemical', aminotriazole. Analytical methods are lacking, too, for certain fungicides in common use for the treatment of seeds – seeds which, if unused at the end of the planting season, may very well find their way into human food.

In effect, then, to establish tolerances is to authorize contamination of public food supplies with poisonous chemicals in order that the farmer and the processor may enjoy the benefit of cheaper production – then to penalize the consumer by taxing him to maintain a policing agency to make certain that he shall not get a lethal dose. But to do the policing job properly would cost money beyond any legislator's courage to appropriate, given the present volume and toxicity of agricultural chemicals. So in the end the luckless consumer pays his taxes but gets his poisons regardless.

What is the solution? The first necessity is the elimination of tolerances on the chlorinated hydrocarbons, the organic phosphorus group, and other highly toxic chemicals. It will immediately be objected that this will place an intolerable burden on the farmer. But if, as is now the presumable goal, it is possible to use chemicals in such a way that they leave a residue of only 7 parts per million (the tolerance for DDT), or of 1 part per million (the tolerance for parathion), or even of only 0·1 part per million as is required for dieldrin on a great variety of fruits and vegetables, then why is it not possible, with only a little more care, to prevent the occurrence of any residues at all? This, in fact, is what is required for some chemicals such as heptachlor, endrin, and dieldrin on certain crops. If it is considered practical in these instances, why not for all?

But this is not a complete or final solution, for a zero tolerance on paper is of little value. At present, as we have seen, more than 99 per cent of the inter-state food shipments slip by without inspection. A vigilant and aggressive Food and Drug Administration, with a greatly increased force of inspectors, is another urgent need.

This system, however – deliberately poisoning our food, then policing the result – is too reminiscent of Lewis Carroll's White Knight who thought of 'a plan to dye one's whiskers green, and always use so large a fan that they could not be seen'. The ultimate answer is to use less toxic chemicals so that the public hazard from their misuse is greatly reduced. Such chemicals already exist: the pyrethrins, rotenone, ryania, and others derived from plant substances. Synthetic substitutes for the pyrethrins have recently been developed so that an otherwise critical shortage can be averted. Public education as to the nature of the chemicals offered for sale is sadly needed. The average purchaser is completely bewildered by the array of available insecticides, fungicides, and weed killers, and has no way of knowing which are the deadly ones, which reasonably safe.

In addition to making this change to less dangerous agricultural pesticides, we should diligently explore the possibilities of non-chemical methods. Agricultural use of insect diseases, caused by a bacterium highly specific for certain types of insects, is already being tried in California, and more extended tests of this method are under way. A great many other possibilities exist for effective insect control by methods that will leave no residues on foods (see Chapter 17). Until a large-scale conversion to these methods has been made, we shall have little relief from a situation that, by any common-sense standards, is intolerable. As matters stand now, we are in little better position than the guests of the Borgias.

Chapter 12

The Human Price

As the tide of chemicals born of the Industrial Age has arisen to engulf our environment, a drastic change has come about in the nature of the most serious public health problems. Only yesterday mankind lived in fear of the scourges of smallpox, cholera, and plague that once swept nations before them. Now our major concern is no longer with the disease organisms that once were omnipresent; sanitation, better living conditions, and new drugs have given us a high degree of control over infectious disease. Today we are concerned with a different kind of hazard that lurks in our environment – a hazard we ourselves have introduced into our world as our modern way of life has evolved.

The new environmental health problems are multiple – created by radiation in all its forms, born of the never-ending stream of chemicals of which pesticides are a part, chemicals now pervading the world in which we live, acting upon us directly and indirectly, separately and collectively. Their presence casts a shadow that is no less ominous because it is formless and obscure, no less frightening because it is simply impossible to predict the effects of lifetime exposure to chemical and physical agents that are not part of the biological experience of man.

We all live under the haunting fear that something may corrupt the environment to the point where man joins the dinosaurs as an obsolete form of life [says Dr David Price of the United States Public Health Service]. And what makes these thoughts all the more disturbing is the knowledge that our fate could perhaps be sealed twenty or more years before the development of symptoms.

Where do pesticides fit into the picture of environmental disease? We have seen that they now contaminate soil, water, and food, that they have the power to make our streams fishless and our gardens and woodlands silent and birdless. Man, however much he may like to pretend the contrary, is part of nature.

Can he escape a pollution that is now so thoroughly distributed throughout our world?

We know that even single exposures to these chemicals, if the amount is large enough, can precipitate acute poisoning. But this is not the major problem. The sudden illness or death of farmers, spraymen, pilots, and others exposed to appreciable quantities of pesticides is tragic and should not occur. For the population as a whole, we must be more concerned with the delayed effects of absorbing small amounts of the pesticides that invisibly contaminate our world.

Responsible public health officials have pointed out that the biological effects of chemicals are cumulative over long periods of time, and that the hazard to the individual may depend on the sum of the exposures received throughout his lifetime. For these very reasons the danger is easily ignored. It is human nature to shrug off what may seem to us a vague threat of future disaster. 'Men are naturally most impressed by diseases which have obvious manifestations,' says a wise physician, Dr René Dubos, 'yet some of their worst enemies creep on them unobtrusively.'

For each of us, as for the robin in Michigan or the salmon in the Miramichi, this is a problem of ecology, of interrelationships, of interdependence. We poison the caddis flies in a stream and the salmon runs dwindle and die. We poison the gnats in a lake and the poison travels from link to link of the food chain and soon the birds of the lake margins become its victims. We spray our elms and the following springs are silent of robin song, not because we sprayed the robins directly but because the poison travelled, step by step, through the now familiar elm leaf-earthworm-robin cycle. These are matters of record, observable, part of the visible world around us. They reflect the web of life – or death – that scientists know as ecology.

But there is also an ecology of the world within our bodies. In this unseen world minute causes produce mighty effects; the effect, moreover, is often seemingly unrelated to the cause, appearing in a part of the body remote from the area where the original injury was sustained. 'A change at one point, in one molecule even, may reverberate throughout the entire system to initiate changes in seemingly unrelated organs and tissues,' says a recent summary of the present status of medical research. When one is concerned with the mysterious and wonderful functioning of the human body, cause and effect are seldom simple and easily demonstrated relationships. They may be widely separated both in space and time. To discover the agent of disease and death depends on a patient piecing together of many seemingly distinct and unrelated facts developed through a vast amount of research in widely separated fields.

We are accustomed to look for the gross and immediate effect and to ignore all else. Unless this appears promptly and in such obvious form that it cannot be ignored, we deny the existence of hazard. Even research men suffer from the handicap of inadequate methods of detecting the beginnings of injury. The lack of sufficiently delicate methods to detect injury before symptoms appear is one of the great unsolved problems in medicine.

'But,' someone will object, 'I have used dieldrin sprays on the lawn many times but I have never had convulsions like the World Health Organization spraymen – so it hasn't harmed me.' It is not that simple. Despite the absence of sudden and dramatic symptoms, one who handles such materials is unquestionably storing up toxic materials in his body. Storage of the chlorinated hydrocarbons, as we have seen, is cumulative, beginning with the smallest intake. The toxic materials become lodged in all the fatty tissues of the body. When these reserves of fat are drawn upon the poison may then strike quickly. A New Zealand medical journal recently provided an example. A man under treatment for obesity suddenly developed symptoms of poisoning. On examination his fat was found to contain stored dieldrin, which had been metabolized as he lost weight. The same thing could happen with loss of weight in illness.

The results of storage, on the other hand, could be even less obvious. Several years ago the *Journal* of the American Medical Association warned strongly of the hazards of insecticide storage in adipose tissue, pointing out that drugs or chemicals that are cumulative require greater caution than those having no tendency to be stored in the tissues. The adipose tissue, we are warned, is not merely a place for the deposition of fat (which makes up about 18 per cent of the body weight), but has many important functions with which the stored poisons may interfere.

Furthermore, fats are very widely distributed in the organs and tissues of the whole body, even being constituents of cell membranes. It is important to remember, therefore, that the fat-soluble insecticides become stored in individual cells, where they are in position to interfere with the most vital and necessary functions of oxidation and energy production. This important aspect of the problem will be taken up in the next chapter.

One of the most significant facts about the chlorinated hydrocarbon insecticides is their effect on the liver. Of all organs in the body the liver is most extraordinary. In its versatility and in the indispensable nature of its functions it has no equal. It presides over so many vital activities that even the slightest damage to it is fraught with serious consequences. Not only does it provide bile for the digestion of fats, but because of its location and the special circulatory pathways that converge upon it the liver receives blood directly from the digestive tract and is deeply involved in the metabolism of all the principal foodstuffs. It stores sugar in the form of glycogen and releases it as glucose in carefully measured quantities to keep the blood sugar at a normal level. It builds body proteins, including some essential elements of blood plasma concerned with blood-clotting. It maintains cholesterol at its proper level in the blood plasma, and inactivates the male and female hormones when they reach excessive levels. It is a storehouse of many vitamins, some of which in turn contribute to its own proper functioning.

Without a normally functioning liver the body would be disarmed – defenceless against the great variety of poisons that continually invade it. Some of these are normal by-products of metabolism, which the liver swiftly and efficiently makes harmless by withdrawing their nitrogen. But poisons that have no normal place in the body may also be detoxified. The 'harmless' insecticides malathion and methoxychlor are less poisonous than their relatives only because a liver enzyme deals with them, altering their molecules in such a way that their capacity for harm is lessened. In similar ways the liver deals with the majority of the toxic materials to which we are exposed.

Our line of defence against invading poisons or poisons from within is now weakened and crumbling. A liver damaged by pesticides is not only incapable of protecting us from poisons, the whole wide range of its activities may be interfered with. Not only are the consequences far-reaching, but because of their variety and the fact that they may not immediately appear they may not be attributed to their true cause.

In connexion with the nearly universal use of insecticides that are liver poisons, it is interesting to note the sharp rise in hepatitis that began during the 1950s and is continuing a fluctuating climb. Cirrhosis also is said to be increasing. While it is admittedly difficult, in dealing with human beings rather than laboratory animals, to 'prove' that cause A produces effect B, plain common sense suggests that the relation between a soaring rate of liver disease and the prevalence of liver poisons in the environment is no coincidence. Whether or not the chlorinated hydrocarbons are the primary cause, it seems hardly sensible under the circumstances to expose ourselves to poisons that have a proven ability to damage the liver and so presumably to make it less resistant to disease.

Both major types of insecticides, the chlorinated hydrocarbons and the organic phosphates, directly affect the nervous system, although in somewhat different ways. This has been made clear by an infinite number of experiments on animals and by observations on human subjects as well. As for DDT, the first of the new organic insecticides to be widely used, its action is primarily on the central nervous system of man; the cerebellum and the higher motor cortex are thought to be the areas chiefly affected. Abnormal sensations as of prickling, burning, or itching, as well as tremors or even convulsions, may follow exposure to appreciable amounts, according to a standard textbook of toxicology.

Our first knowledge of the symptoms of acute poisoning by DDT was furnished by several British investigators, who deliberately exposed themselves in order to learn the consequences. Two scientists at the British Royal Navy Physiological Laboratory invited absorption of DDT through the skin by direct contact with walls covered with a water-soluble paint containing 2 per cent DDT, overlaid with a thin film of oil. The direct effect on the nervous system is apparent in their eloquent description of their symptoms:

The tiredness, heaviness, and aching of limbs were very real things, and the mental state was also most distressing . . . [there was] extreme irritability... great distaste for work of any sort... a feeling of mental incompetence in tackling the simplest mental task. The joint pains were quite violent at times.

Another British experimenter who applied DDT in acetone solution to his skin reported heaviness and aching of limbs, muscular weakness, and 'spasms of extreme nervous tension'. He took a holiday and improved, but on return to work his condition deteriorated. He then spent three weeks in bed, made miserable by constant aching in limbs, insomnia, nervous tension, and feelings of acute anxiety. On occasion tremors shook his whole body – tremors of the sort now made all too familiar by the sight of birds poisoned by DDT. The experimenter lost ten weeks from his work, and at the end of a year, when his case was reported in a British medical journal, recovery was not complete.

(Despite this evidence, several American investigators conducting an experiment with DDT on volunteer subjects dismissed the complaint of headache and 'pain in every bone' as 'obviously of psychoneurotic origin'.)

There are now many cases on record in which both the symptoms and the whole course of the illness point to insecticides as the cause. Typically, such a victim has had a known exposure to one of the insecticides, his symptoms have subsided under treatment which included the exclusion of all insecticides from his environment, and most significantly *have returned with each renewed contact* with the offending chemicals. This sort of evidence – and no more – forms the basis of a vast amount of medical therapy in many other disorders. There is no reason why it should not serve as a warning that it is no longer sensible to take the 'calculated risk' of saturating our environment with pesticides.

Why does not everyone handling and using insecticides develop the same symptoms? Here the matter of individual sensitivity enters in. There is some evidence that women are more susceptible than men, the very young more than adults, those who lead sedentary, indoor lives more than those leading a rugged life of work or exercise in the open. Beyond these differences are others that are no less real because they are intangible. What makes one person allergic to dust or pollen, sensitive to a poison, or susceptible to an infection whereas another is not is a medical mystery for which there is at present no explanation. The problem nevertheless exists and it affects significant numbers of the population. Some physicians estimate that a third or more of their patients show signs of some form of sensitivity, and that the number is growing. And unfortunately, sensitivity may suddenly develop in a person previously insensitive. In fact, some medical men believe that intermittent exposures to chemicals may produce just such sensitivity. If this is true, it may explain why some studies on men subjected to continuous occupational exposure find little evidence of toxic effects. By their constant contact with the chemicals these men keep themselves desensitized – as an allergist keeps his patients desensitized by repeated small injections of the allergen.

The whole problem of pesticide poisoning is enormously complicated by the fact that a human being, unlike a laboratory animal living under rigidly controlled conditions, is never exposed to one chemical alone. Between the major groups of insecticides, and between them and other chemicals, there are interactions that have serious potentials. Whether released into soil or water or a man's blood, these unrelated chemicals do not remain segregated; there are mysterious and unseen changes by which one alters the power of another for harm.

There is interaction even between the two major groups of insecticides usually thought to be completely distinct in their action. The power of the organic phosphates, those poisoners of the nerve-protective enzyme cholinesterase, may become greater if the body has first been exposed to a chlorinated hydrocarbon which injures the liver. This is because, when liver function is disturbed, the cholinesterase level drops below normal. The added depressive effect of the organic phosphate may then be enough to precipitate acute symptoms. And as we have seen, pairs of the organic phosphates themselves may interact in such a way as to increase their toxicity a hundredfold. Or the organic phosphates may interact with various drugs, or with synthetic materials, food additives – who can say what else of the infinite number of man-made substances that now pervade our world?

The effect of a chemical of supposedly innocuous nature can be drastically changed by the action of another; one of the best examples is a close relative of DDT called methoxychlor. (Actually, methoxychlor may not be as free from dangerous qualities as it is generally said to be, for recent work on experimental animals shows a direct action on the uterus and a blocking effect on some of the powerful pituitary hormones – reminding us again that these are chemicals with enormous biologic effect. Other work shows that methoxychlor has a potential ability to damage the kidneys.) Because it is not stored to any great extent when given alone, we are told that methoxychlor is a safe chemical. But this is not necessarily true. If the liver has been damaged by another agent, methoxychlor is stored in the body at *100 times* its normal rate, and will then imitate the effects of DDT with long-lasting effects on the nervous system. Yet the liver damage that brings this about might be so slight as to pass unnoticed. It might have been the result of any of a number of commonplace situations – using another insecticide, using a cleaning fluid containing carbon tetrachloride, or taking one of the so-called tranquillizing drugs, a number (but not all) of which are chlorinated hydrocarbons and possess power to damage the liver.

Damage to the nervous system is not confined to acute poisoning; there may also be delayed effects from exposure. Long-lasting damage to brain or nerves has been reported for methoxychlor and others. Dieldrin, besides its immediate consequences, can have long-delayed effects ranging from 'loss of memory, insomnia, and nightmares to mania'. Lindane, according to medical findings, is stored in significant amounts in the brain and functioning liver tissue and may induce 'profound and long-lasting effects on the central nervous system'. Yet this chemical, a form of benzene hexachloride, is much used in vaporizers, devices that pour a stream of volatilized insecticide vapour into homes, offices, restaurants.

The organic phosphates, usually considered only in relation to their more violent manifestations in acute poisoning, also have the power to produce lasting physical damage to nerve tissues and, according to recent findings, to induce mental disorders. Various cases of delayed paralysis have followed use of one or another of these insecticides. A bizarre happening in the United States during the prohibition era about 1930 was an omen of things to come. It was caused not by an insecticide but by a substance belonging chemically to the same group as the organic phosphate insecticides. During that period some medicinal substances were being pressed into service as substitutes for liquor, being exempt from the prohibition law. One of these was Jamaica ginger. But the *United States Pharmacopeia* product was expensive, and bootleggers conceived the idea of making a substitute Jamaica ginger. They succeeded so well that their spurious product responded to the appropriate chemical tests and deceived the government chemists. To give their false ginger the necessary tang they had introduced a chemical known as triorthocresyl phosphate. This chemical, like parathion and its relatives, destroys the protective enzyme cholinesterase. As a consequence of drinking the bootleggers' product some 15,000 people developed a permanently crippling type of paralysis of the leg muscles, a condition now called 'ginger paralysis'. The paralysis was accompanied by destruction of the nerve sheaths and by degeneration of the cells of the anterior horns of the spinal cord.

About two decades later various other organic phosphates came into use as insecticides, as we have seen, and soon cases reminiscent of the ginger paralysis episode began to occur. One was a greenhouse worker in Germany who became paralysed several months after experiencing mild symptoms of poisoning on a few occasions after using parathion. Then a group of three chemical plant workers developed acute poisoning from exposure to other insecticides of this group. They recovered under treatment, but ten days later two of them developed muscular weakness in the legs. This persisted for ten months in one; the other, a young woman chemist, was more severely affected, with paralysis in both legs and some involvement of the hands and arms. Two years later when her case was reported in a medical journal she was still unable to walk.

The insecticide responsible for these cases has been withdrawn from the market, but some of those now in use may be capable of like harm. Malathion (beloved of gardeners) has induced severe muscular weakness in experiments on chickens. This was attended (as in ginger paralysis) by destruction of the sheaths of the sciatic and spinal nerves.

All these consequences of organic phosphate poisoning, if survived, may be a prelude to worse. In view of the severe damage they inflict upon the nervous system, it was perhaps inevitable that these insecticides would eventually be linked with mental disease. That link has recently been supplied by investigators at the University of Melbourne and Prince Henry's Hospital in Melbourne, who reported on sixteen cases of mental disease. All had a history of prolonged exposure to organic phosphorus insecticides. Three were scientists checking the efficacy of sprays; eight worked in greenhouses; five were farm workers. Their symptoms ranged from impairment of memory to schizophrenic and depressive reactions. All had normal medical histories before the chemicals they were using boomeranged and struck them down.

Echoes of this sort of thing are to be found, as we have seen, widely scattered throughout medical literature, sometimes involving the chlorinated hydrocarbons, sometimes the organic phosphates. Confusion, delusions, loss of memory, mania – a heavy price to pay for the temporary destruction of a few insects, but a price that will continue to be exacted as long as we insist upon using chemicals that strike directly at the nervous system.

Chapter 13

Through a Narrow Window

The biologist George Wald once compared his work on an exceedingly specialized subject, the visual pigments of the eye, to a very narrow window through which at a distance one can see only a crack of light. As one comes closer the view grows wider and wider, until finally through this same narrow window one is looking at the universe.

So it is that only when we bring our focus to bear, first on the individual cells of the body, then on the minute structures within the cells, and finally on the ultimate reactions of molecules within these structures – only when we do this can we comprehend the most serious and far-reaching effects of the haphazard introduction of foreign chemicals into our internal environment. Medical research has only rather recently turned to the functioning of the individual cell in producing the energy that is the indispensable quality of life. The extraordinary energy-producing mechanism of the body is basic not only to health but to life; it transcends in importance even the most vital organs, for without the smooth and effective functioning of energy-yielding oxidation none of the body's functions can be performed. Yet the nature of many of the chemicals used against insects, rodents, and weeds is such that they may strike directly at this system, disrupting its beautifully functioning mechanism.

The research that led to our present understanding of cellular oxidation is one of the most impressive accomplishments in all biology and biochemistry. The roster of contributors to this work includes many Nobel Prize winners. Step by step it has been going on for a quarter of a century, drawing on even earlier work for some of its foundation stones. Even yet it is not complete in all details. And only within the past decade have all the varied pieces of research come to form a whole so that biological oxidation could become part of the common knowledge of biologists. Even more important is the fact that medical men who received their basic training before 1950 have had little opportunity to realize the critical importance of the process and the hazards of disrupting it.

The ultimate work of energy production is accomplished not in any specialized organ but in every cell of the body. A living cell, like a flame, burns fuel to produce the energy on which life depends. The analogy is more poetic than precise, for the cell accomplishes its 'burning' with only the moderate heat of the body's normal temperature. Yet all these billions of gently burning little fires spark the energy of life. Should they cease to burn,

no heart could beat, no plant could grow upward defying gravity, no amoeba could swim, no sensation could speed along a nerve, no thought could flash in the human brain,

said the chemist Eugene Rabinowitch.

The transformation of matter into energy in the cell is an ever-flowing process, one of nature's cycles of renewal, like a wheel endlessly turning. Grain by grain, molecule by molecule, carbohydrate fuel in the form of glucose is fed into this wheel; in its cyclic passage the fuel molecule undergoes fragmentation and a series of minute chemical changes. The changes are made in orderly fashion, step by step, each step directed and controlled by an enzyme of so specialized a function that it does this one thing and nothing else. At each step energy is produced, waste products (carbon dioxide and water) are given off, and the altered molecule of fuel is passed on to the next stage. When the turning wheel comes full cycle the fuel molecule has been stripped down to a form in which it is ready to combine with a new molecule coming in and to start the cycle anew.

This process by which the cell functions as a chemical factory is one of the wonders of the living world. The fact that all the functioning parts are of infinitesimal size adds to the miracle. With few exceptions cells themselves are minute, seen only with the aid of a microscope. Yet the greater part of the work of oxidation is performed in a theatre far smaller, in tiny granules within the cell called mitochondria. Although known for more than sixty years, these were formerly dismissed as cellular elements of unknown and probably unimportant function. Only in the 1950s did their study become an exciting and fruitful field of research; suddenly they began to engage so much attention that a thousand papers on this subject alone appeared within a five-year period.

Again one stands in awe at the marvellous ingenuity and patience by which the mystery of the mitochondria has been solved. Imagine a particle so small that you can barely see it even though a microscope has enlarged it for you 300 times. Then imagine the skill required to isolate this particle, to take it apart and analyse its components and determine their highly complex functioning. Yet this has been done with the aid of the electron microscope and the techniques of the biochemist.

It is now known that the mitochondria are tiny packets of enzymes, a varied assortment including all the enzymes necessary for the oxidative cycle, arranged in precise and orderly array on walls and partitions. The mitochondria are the 'powerhouses' in which most of the energy-producing reactions occur. After the first, preliminary steps of oxidation have been performed in the cytoplasm the fuel molecule is taken into the mitochondria. It is here that oxidation is completed; it is here that enormous amounts of energy are released.

The endlessly turning wheels of oxidation within the mitochondria would turn to little purpose if it were not for this all-important result. The energy produced at each stage of the oxidative cycle is in a form familiarly spoken of by the biochemists as ATP (adenosine triphosphate), a molecule containing three phosphate groups. The role of ATP in furnishing energy comes from the fact that it can transfer one of its phosphate groups to other substances, along with the energy of its bonds of electrons shuttling back and forth at high speed. Thus, in a muscle cell, energy to contract is gained when a terminal phosphate group is transferred to the contracting muscle. So another cycle takes place – a cycle within a cycle: a molecule of ATP gives up one of its phosphate groups and retains only two, becoming a diphosphate molecule, ADP. But as the wheel turns further another phosphate group is coupled on and the potent ATP is restored. The analogy of the storage battery has been used: ATP represents the charged, ADP the discharged battery.

ATP is the universal currency of energy – found in all organisms from microbes to man. It furnishes mechanical energy to muscle cells; electrical energy to nerve cells. The sperm cell, the fertilized egg ready for the enormous burst of activity that will transform it into a frog or a bird or a human infant, the cell that must create a hormone, all are supplied with ATP. Some of the energy of ATP is used in the mitochondrion but most of it is immediately dispatched into the cell to provide power for other activities. The location of the mitochondria within certain cells is eloquent of their function, since they are placed so that energy can be delivered precisely where it is needed. In muscle cells they cluster around contracting fibres; in nerve cells they are found at the junction with another cell, supplying energy for the transfer of impulses; in sperm cells they are concentrated at the point where the propellant tail is joined to the head.

The charging of the battery, in which ADP and a free phosphate group are combined to restore ATP, is coupled to the oxidative process; the close linking is known as coupled phosphorylation. If the combination becomes uncoupled, the means is lost for providing usable energy. Respiration continues but no energy is produced. The cell has become like a racing engine, generating heat but yielding no power. Then the muscle cannot contract, nor can the impulse race along the nerve pathways. Then the sperm cannot move to its destination; the fertilized egg cannot carry to completion its complex divisions and elaborations. The consequences of uncoupling could indeed be disastrous for any organism from embryo to adult: in time it could lead to the death of the tissue or even of the organism.

How can uncoupling be brought about? Radiation is an uncoupler, and the death of cells exposed to radiation is thought by some to be brought about in this way. Unfortunately, a good many chemicals also have the power to separate oxidation from energy production, and the insecticides and weed killers are well represented on the list. The phenols, as we have seen, have a strong effect on metabolism, causing a potentially fatal rise in temperature; this is brought about by the 'racing engine' effect of uncoupling. The dinitrophenols and pentachlorophenols are examples of this group that have widespread use as herbicides. Another uncoupler among the herbicides is 2,4-D. Of the chlorinated hydrocarbons, DDT is a proven uncoupler and further study will probably reveal others among this group.

But uncoupling is not the only way to extinguish the little fires in some or all of the body's billions of cells. We have seen that each step in oxidation is directed and expedited by a specific enzyme. When any of these enzymes – even a single one of them – is destroyed or weakened, the cycle of oxidation within the cell comes to a halt. It makes no difference which enzyme is affected. Oxidation progresses in a cycle like a turning wheel. If we thrust a crowbar between the spokes of a wheel it makes no difference where we do it, the wheel stops turning. In the same way, if we destroy an enzyme that functions at any point in the cycle, oxidation ceases. There is then no further energy production, so the end effect is very similar to uncoupling.

The crowbar to wreck the wheels of oxidation can be supplied by any of a number of chemicals commonly used as pesticides. DDT, methoxychlor, malathion, phenothiazine, and various dinitro compounds are among the numerous pesticides that have been found to inhibit one or more of the enzymes concerned in the cycle of oxidation. They thus appear as agents potentially capable of blocking the whole process of energy production and depriving the cells of utilizable oxygen. This is an injury with most disastrous consequences, only a few of which can be mentioned here.

Merely by systematically withholding oxygen, experimenters have caused normal cells to turn into cancer cells, as we shall see in the following chapter. Some hint of other drastic consequences of depriving a cell of oxygen can be seen in animal experiments on developing embryos. With insufficient oxygen the orderly processes by which the tissues unfold and the organs develop are disrupted; malformations and other abnormalities then occur. Presumably the human embryo deprived of oxygen may also develop congenital deformities.

There are signs that an increase in such disasters is being noticed, even though few look far enough to find all of the causes. In one of the more unpleasant portents of the times, the Office of Vital Statistics in 1961 initiated a national tabulation of malformations at birth, with the explanatory comment that the resulting statistics would provide needed facts on the incidence of congenital malformations and the circumstances under which they occur. Such studies will no doubt be directed largely towards measuring the effects of radiation, but it must not be overlooked that many chemicals are the partners of radiation, producing precisely the same effects. Some of the defects and malformations in tomorrow's children, grimly anticipated by the Office of Vital Statistics, will almost certainly be caused by these chemicals that permeate our outer and inner worlds.

It may well be that some of the findings about diminished reproduction are also linked with interference with biological oxidation, and consequent depletion of the all-important storage batteries of ATP. The egg, even before fertilization, needs to be generously supplied with ATP, ready and waiting for the enormous effort, the vast expenditure of energy that will be required once the sperm has entered and fertilization has occurred. Whether the sperm cell will reach and penetrate the egg depends upon its own supply of ATP, generated in the mitochondria thickly clustered in the neck of the cell. Once fertilization is accomplished and cell division has begun, the supply of energy in the form of ATP will largely determine whether the development of the embryo will proceed to completion. Embryologists studying some of their most convenient subjects, the eggs of frogs and of sea urchins, have found that if the ATP content is reduced below a certain critical level the egg simply stops dividing and soon dies.

It is not an impossible step from the embryology laboratory to the apple tree where a robin's nest holds its complement of blue-green eggs; but the eggs lie cold, the fires of life that flickered for a few days now extinguished. Or to the top of a tall Florida pine where a vast pile of twigs and sticks in ordered disorder holds three large white eggs, cold and lifeless. Why did the robins and the eaglets not hatch? Did the eggs of the birds, like those of the laboratory frogs, stop developing simply because they lacked enough of the common currency of energy – the ATP molecules – to complete their development? And was the lack of ATP brought about because in the body of the parent birds and in the eggs there were stored enough insecticides to stop the little turning wheels of oxidation on which the supply of energy depends?

It is no longer necessary to guess about the storage of insecticides in the eggs of birds, which obviously lend themselves to this kind of observation more readily than the mammalian ovum. Large residues of DDT and other hydrocarbons have been found whenever looked for in the eggs of birds subjected to these chemicals, either experimentally or in the wild. And the concentrations have been heavy. Pheasant eggs in a California experiment contained up to 349 parts per million of DDT. In Michigan, eggs taken from the oviducts of robins dead of DDT poisoning showed concentrations up to 200 parts per million. Other eggs were taken from nests left unattended as parent robins were stricken with poison; these too contained DDT. Chickens poisoned by aldrin used on a neighbouring farm have passed on the chemical to their eggs; hens experimentally fed DDT laid eggs containing as much as 65 parts per million.

Knowing that DDT and other (perhaps all) chlorinated hydrocarbons stop the energy-producing cycle by inactivating a specific enzyme or uncoupling the energy-producing mechanism, it is hard to see how any egg so loaded with residues could complete the complex process of development: the infinite number of cell divisions, the elaboration of tissues and organs, the synthesis of vital substances that in the end produce a living creature. All this requires vast amounts of energy – the little packets of ATP which the turning of the metabolic wheel alone can produce.

There is no reason to suppose these disastrous events are confined to birds. ATP is the universal currency of energy, and the metabolic cycles that produce it turn to the same purpose in birds and bacteria, in men and mice. The fact of insecticide storage in the germ cells of any species should therefore disturb us, suggesting comparable effects in human beings.

And there are indications that these chemicals lodge in tissues concerned with the manufacture of germ cells as well as in the cells themselves. Accumulations of insecticides have been discovered in the sex organs of a variety of birds and mammals – in pheasants, mice, and guinea-pigs under controlled conditions, in robins in an area sprayed for elm disease, and in deer roaming western forests sprayed for spruce budworm. In one of the robins the concentration of DDT in the testes was heavier than in any other part of the body. Pheasants also accumulated extraordinary amounts in the testes, up to 1,500 parts per million.

Probably as an effect of such storage in the sex organs, atrophy of the testes has been observed in experimental mammals. Young rats exposed to methoxychlor had extraordinarily small testes. When young roosters were fed DDT, the testes made only 18 per cent of their normal growth; combs and wattles, dependent for their development upon the testicular hormone, were only a third the normal size.

The spermatozoa themselves may well be affected by loss of ATP. Experiments show that the motility of bull sperm is decreased by dinitrophenol, which interferes with the energy-coupling mechanism with inevitable loss of energy. The same effect would probably be found with other chemicals were the matter investigated. Some indication of the possible effect on human beings is seen in medical reports of oligospermia, or reduced production of spermatozoa, among aviation crop dusters applying DDT.

For mankind as a whole, a possession infinitely more valuable than individual life is our genetic heritage, our link with past and future. Shaped through long aeons of evolution, our genes not only make us what we are, but hold in their minute beings the future – be it one of promise or threat. Yet genetic deterioration through man-made agents is the menace of our time, 'the last and greatest danger to our civilization'.

Again the parallel between chemicals and radiation is exact and inescapable.

The living cell assaulted by radiation suffers a variety of injuries: its ability to divide normally may be destroyed, it may suffer changes in chromosome structure, or the genes, carriers of hereditary material, may undergo those sudden changes known as mutations, which cause them to produce new characteristics in succeeding generations. If especially susceptible the cell may be killed outright, or finally, after the passage of time measured in years, it may become malignant.

All these consequences of radiation have been duplicated in laboratory studies by a large group of chemicals known as radio-mimetic or radiation-imitating. Many chemicals used as pesticides – herbicides as well as insecticides – belong to this group of substances that have the ability to damage the chromosomes, interfere with normal cell division, or cause mutations. These injuries to the genetic material are of a kind that may lead to disease in the individual exposed or they may make their effects felt in future generations.

Only a few decades ago, no one knew these effects of either radiation or chemicals. In those days the atom had not been split and few of the chemicals that were to duplicate radiation had as yet been conceived in the test tubes of chemists. Then in 1927, a professor of zoology in a Texas university, Dr H. J. Muller, found that by exposing an organism to X-radiation, he could produce mutations in succeeding generations. With Muller's discovery a vast new field of scientific and medical knowledge was opened up. Muller later received the Nobel Prize in Medicine for his achievement, and in a world that soon gained unhappy familiarity with the grey rains of fallout, even the non-scientist now knows the potential results of radiation.

Although far less noticed, a companion discovery was made by Charlotte Auerbach and William Robson at the University of Edinburgh in the early 1940s. Working with mustard gas, they found that this chemical produces permanent chromosome abnormalities that cannot be distinguished from those induced by radiation. Tested on the fruit fly, the same organism Muller had used in his original work with X-rays, mustard gas also produced mutations. Thus the first chemical mutagen was discovered.

Mustard gas as a mutagen has now been joined by a long list of other chemicals known to alter genetic material in plants and animals. To understand how chemicals can alter the course of heredity, we must first watch the basic drama of life as it is played on the stage of the living cell.

The cells composing the tissues and organs of the body must have the power to increase in number if the body is to grow and if the stream of life is to be kept flowing from generation to generation. This is accomplished by the process of mitosis, or nuclear division. In a cell that is about to divide, changes of the utmost importance occur, first within the nucleus, but eventually involving the entire cell. Within the nucleus, the chromosomes mysteriously move and divide, ranging themselves in age-old patterns that will serve to distribute the determiners of heredity, the genes, to the daughter cells. First they assume the form of elongated threads, on which the genes are aligned, like beads on a string. Then each chromosome divides lengthwise (the genes dividing also). When the cell divides into two, half of each goes to each of the daughter cells. In this way each new cell will contain a complete set of chromosomes, and all the genetic information encoded within them. In this way the integrity of the race and of the species is preserved; in this way like begets like.

A special kind of cell division occurs in the formation of the germ cells. Because the chromosome number for a given species is constant, the egg and the sperm, which are to unite to form a new individual, must carry to their union only half the specific number. This is accomplished with extraordinary precision by a change in the behaviour of the chromosomes that occurs at one of the divisions producing those cells. At this time the chromosomes do not split, but one whole chromosome of each pair goes into each daughter cell.

In this elemental drama all life is revealed as one. The events of the process of cell division are common to all earthly life; neither man nor amoeba, the giant sequoia nor the simple yeast cell can long exist without carrying on this process of cell division. Anything that disturbs mitosis is therefore a grave threat to the welfare of the organism affected and to its descendants.

The major features of cellular organization, including, for instance, mitosis, must be much older than 500 million years – more nearly 1,000 million [wrote George Gaylord Simpson and his colleagues Pittendrigh and Tiffany in their broadly encompassing book entitled *Life*]. In this sense the world of life, while surely fragile and complex, is incredibly durable through time – more durable than mountains. This durability is wholly dependent on the almost incredible accuracy with which the inherited information is copied from generation to generation.

But in all the thousand million years envisioned by these authors no threat has struck so directly and so forcefully at that 'incredible accuracy' as the mid-twentieth-century threat of man-made radiation and man-made and man-disseminated chemicals. Sir Macfarlane Burnet, a distinguished Australian physician and a Nobel Prize winner, considers it

one of the most significant medical features [of our time that] as a by-product of more and more powerful therapeutic procedures and the production of chemical substances outside of biological experiences, the normal protective barriers that kept mutagenic agents from the internal organs have been more and more frequently penetrated.

The study of human chromosomes is in its infancy, and so it has only recently become possible to study the effect of environmental factors upon them. It was not until 1956 that new techniques made it possible to determine accurately the number of chromosomes in the human cell – forty-six – and to observe them in such detail that the presence or absence of whole chromosomes or even parts of chromosomes could be detected. The whole concept of genetic damage by something in the environment is also relatively new, and is little understood except by the geneticist, whose advice is too seldom sought. The hazard from radiation in its various forms is now reasonably well understood – although still denied in surprising places. Dr Muller has frequently had occasion to deplore the

resistance to the acceptance of genetic principles on the part of so many, not only of governmental appointees in the policy-making positions, but also of so many of the medical profession.

The fact that chemicals may play a role similar to radiation has scarcely dawned on the public mind, nor on the minds of most medical or scientific workers. For this reason the role of chemicals in general use (rather than in laboratory experiments) has not yet been assessed. It is extremely important that this be done.

Sir Macfarlane is not alone in his estimate of the potential danger. Dr Peter Alexander, an outstanding British authority, has said that the radio-mimetic chemicals 'may well represent a greater danger' than radiation. Dr Muller, with the perspective gained by decades of distinguished work in genetics, warns that various chemicals (including groups represented by pesticides)

can raise the mutation frequency as much as radiation.... As yet far too little is known of the extent to which our genes, under modern conditions of exposure to unusual chemicals, are being subjected to such mutagenic influences.

The widespread neglect of the problem of chemical mutagens is perhaps due to the fact that those first discovered were of scientific interest only. Nitrogen mustard, after all, is not sprayed upon whole populations from the air; its use is in the hands of experimental biologists or of physicians who use it in cancer therapy. (A case of chromosome damage in a patient receiving such therapy has recently been reported.) But insecticides and weed killers *are* brought into intimate contact with large numbers of people.

Despite the scant attention that has been given to the matter, it is possible to assemble specific information on a number of these pesticides, showing that they disturb the cell's vital processes in ways ranging from slight chromosome damage to gene mutation, and with consequences extending to the ultimate disaster of malignancy.

Mosquitoes exposed to DDT for several generations turned into strange creatures called gynandromorphs – part male and part female.

Plants treated with various phenols suffered profound destruction of chromosomes, changes in genes, a striking number of mutations, 'irreversible hereditary changes'. Mutations also occurred in fruit flies, the classic subject of genetics experiments, when subjected to phenol; these flies developed mutations so damaging as to be fatal on exposure to one of the common herbicides or to urethane. Urethane belongs to the group of chemicals called carbamates, from which an increasing number of insecticides and other agricultural chemicals are drawn. Two of the carbamates are actually used to prevent sprouting of potatoes in storage – precisely because of their proven effect in stopping cell division. One of these, maleic hydrazide, is rated a powerful mutagen.

Plants treated with benzene hexachloride (BHC) or lindane became monstrously deformed with tumour-like swellings on their roots. Their cells grew in size, being swollen with chromosomes which doubled in number. The doubling continued in future divisions until further cell division became mechanically impossible.

The herbicide 2,4-D has also produced tumour-like swellings in treated plants. Chromosomes become short, thick, clumped together. Cell division is seriously retarded. The general effect is said to parallel closely that produced by X-rays.

These are but a few illustrations, many more could be cited. As yet there has been no comprehensive study aimed at testing the mutagenic effects of pesticides as such. The facts cited above are by-products of research in cell physiology or genetics. What is urgently needed is a direct attack on the problem.

Some scientists who are willing to concede the potent effect of environmental radiation on man nevertheless question whether metagenic chemicals can, as a practical proposition, have the same effect. They cite the great penetrating power of radiation, but doubt that chemicals could reach the germ cells. Once again we are hampered by the fact that there has been little direct investigation of the problem in man. However, the finding of large residues of DDT in the gonads and germ cells of birds and mammals is strong evidence that the chlorinated hydrocarbons, at least, not only become widely distributed throughout the body but come into contact with genetic materials. Professor David E. Davis at Pennsylvania State University has recently discovered that a potent chemical which prevents cells from dividing and has had limited use in cancer therapy can also be used to cause sterility in birds. Sub-lethal levels of the chemical halt cell division in the gonads. Professor Davis has had some success in field trials. Obviously, then, there is little basis for the hope or belief that the gonads of any organism are shielded from chemicals in the environment.

Recent medical findings in the field of chromosome abnormalities are of extreme interest and significance. In 1959 several British and French research teams found their independent studies pointing to a common conclusion – that some of humanity's ills are caused by a disturbance of the normal chromosome number. In certain diseases and abnormalities studied by these investigators the number differed from the normal. To illustrate: it is now known that all typical mongoloids have one extra chromosome. Occasionally this is attached to another so that the chromosome number remains the normal forty-six. As a rule, however, the extra is a separate chromosome, making the number forty-seven. In such individuals, the original cause of the defect must have occurred in the generation preceding its appearance.

A different mechanism seems to operate in a number of patients, both in America and Great Britain, who are suffering from a chronic form of leukaemia. These have been found to have a consistent chromosome abnormality in some of the blood cells. The abnormality consists of the loss of part of a chromosome. In these patients the skin cells have a normal complement of chromosomes. This indicates that the chromosome defect did not occur in the germ cells that gave rise to these individuals, but represents damage to particular cells (in this case, the precursors of blood cells) that occurred during the life of the individual. The loss of part of a chromosome has perhaps deprived these cells of their 'instructions' for normal behaviour.

The list of defects linked to chromosome disturbances has grown with surprising speed since the opening of this territory, hitherto beyond the boundaries of medical research. One, known only as Klinefelter's syndrome, involves a duplication of one of the sex chromosomes. The resulting individual is a male, but because he carries two of the x chromosomes (becoming xxy instead of xy, the normal male complement) he is somewhat abnormal. Excessive height and mental defects often accompany the sterility caused by this condition. In contrast, an individual who receives only one sex chromosome (becoming xo instead of either xx or xy) is actually female but lacks many of the secondary sexual characteristics. The condition is accompanied by various physical (and sometimes mental) defects, for of course the x chromosome carries genes for a variety of characteristics. This is known as Turner's syndrome. Both conditions had been described in medical literature long before the cause was known.

An immense amount of work on the subject of chromosome abnormalities is being done by workers in many countries. A group at the University of Wisconsin, headed by Dr Klaus Patau, has been concentrating on a variety of congenital abnormalities, usually including mental retardation, that seem to result from the duplication of only part of a chromosome, as if somewhere in the formation of one of the germ cells a chromosome had broken and the pieces had not been properly redistributed. Such a mishap is likely to interfere with the normal development of the embryo.

According to present knowledge, the occurrence of an entire

extra body chromosome is usually lethal, preventing survival of the embryo. Only three such conditions are known to be viable; one of them, of course, is mongolism. The presence of an extra attached fragment, on the other hand, although seriously damaging is not necessarily fatal, and according to the Wisconsin investigators this situation may well account for a substantial part of the so far unexplained cases in which a child is born with multiple defects, usually including mental retardation.

This is so new a field of study that as yet scientists have been more concerned with identifying the chromosome abnormalities associated with disease and defective development than with speculating about the causes. It would be foolish to assume that any single agent is responsible for damaging the chromosomes or causing their erratic behaviour during cell division. But can we afford to ignore the fact that we are now filling the environment with chemicals that have the power to strike directly at the chromosomes, affecting them in the precise ways that could cause such conditions? Is this not too high a price to pay for a sproutless potato or a mosquitoless patio?

We can, if we wish, reduce this threat to our genetic heritage, a possession that has come down to us through some two billion years of evolution and selection of living protoplasm, a possession that is ours for the moment only, until we must pass it on to generations to come. We are doing little now to preserve its integrity. Although chemical manufacturers are required by law to test their materials for toxicity, they are not required to make the tests that would reliably demonstrate genetic effect, and they do not do so.

Chapter 14

One in Every Four

The battle of living things against cancer began so long ago that its origin is lost in time. But it must have begun in a natural environment, in which whatever life inhabited the earth was subjected, for good or ill, to influences that had their origin in sun and storm and the ancient nature of the earth. Some of the elements of this environment created hazards to which life had to adjust or perish. The ultra-violet radiation in sunlight could cause malignancy. So could radiations from certain rocks, or arsenic washed out of soil or rocks to contaminate food or water supplies.

The environment contained these hostile elements even before there was life; yet life arose, and over the millions of years it came to exist in infinite numbers and endless variety. Over the aeons of unhurried time that is nature's, life reached an adjustment with destructive forces as selection weeded out the less adaptable and only the most resistant survived. These natural cancer-causing agents are still a factor in producing malignancy; however, they are few in number and they belong to that ancient array of forces to which life has been accustomed from the beginning.

With the advent of man the situation began to change, for man, alone of all forms of life, can *create* cancer-producing substances, which in medical terminology are called carcinogens. A few man-made carcinogens have been part of the environment for centuries. An example is soot, containing aromatic hydrocarbons. With the dawn of the industrial era the world became a place of continuous, ever-accelerating change. Instead of the natural environment there was rapidly substituted an artificial one composed of new chemical and physical agents, many of them possessing powerful capacities for inducing biologic change. Against these carcinogens which his own activities had created man had no protection, for even as his biological heritage has evolved slowly, so it adapts slowly to new conditions. As a result these powerful substances could easily penetrate the inadequate defences of the body.

The history of cancer is long, but our recognition of the agents that produce it has been slow to mature. The first awareness that external or environmental agents could produce malignant change dawned in the mind of a London physician nearly two centuries ago. In 1775 Sir Percival Pott declared that the scrotal cancer so common among chimney sweeps must be caused by the soot that accumulated on their bodies. He could not furnish the 'proof' we would demand today, but modern research methods have now isolated the deadly chemical in soot and proved the correctness of his perception.

For a century or more after Pott's discovery there seems to have been little further realization that certain of the chemicals in the human environment could cause cancer by repeated skin contact, inhalation, or swallowing. True, it had been noticed that skin cancer was prevalent among workers exposed to arsenic fumes in copper smelters and tin foundries in Cornwall and Wales. And it was realized that workers in the cobalt mines in Saxony and in the uranium mines at Joachimsthal in Bohemia were subject to a disease of the lungs, later identified as cancer. But these were phenomena of the pre-industrial era, before the flowering of the industries whose products were to pervade the environment of almost every living thing.

The first recognition of malignancies traceable to the age of industry came during the last quarter of the nineteenth century. About the time that Pasteur was demonstrating the microbial origin of many infectious diseases, others were discovering the chemical origin of cancer – skin cancers among workers in the new lignite industry in Saxony and in the Scottish shale industry, along with other cancers caused by occupational exposure to tar and pitch. By the end of the nineteenth century a half-dozen sources of industrial carcinogens were known; the twentieth century was to create countless new cancer-causing chemicals and to bring the general population into intimate contact with them. In the less than two centuries intervening since the work of Pott, the environmental situation has been vastly changed.

No longer are exposures to dangerous chemicals occupational alone; they have entered the environment of everyone – even of children as yet unborn. It is hardly surprising, therefore, that we are now aware of an alarming increase in malignant disease.

The increase itself is no mere matter of subjective impressions. The monthly report of the Office of Vital Statistics for July 1959 states that malignant growths, including those of the lymphatic and blood-forming tissues, accounted for 15 per cent of the deaths in 1958 compared with only 4 per cent in 1900. Judging by the present incidence of the disease, the American Cancer Society estimates that 45,000,000 Americans now living will eventually develop cancer. This means that malignant disease will strike two out of three families.

The situation with respect to children is even more deeply disturbing. A quarter-century ago, cancer in children was considered a medical rarity. *Today, more American school children die of cancer than from any other disease.* So serious has this situation become that Boston has established the first hospital in the United States devoted exclusively to the treatment of children with cancer. Twelve per cent of all deaths in children between the ages of one and fourteen are caused by cancer. Large numbers of malignant tumours are discovered clinically in children under the age of five, but it is an even grimmer fact that significant numbers of such growths are present at or before birth. Dr W. C. Hueper of the National Cancer Institute, a foremost authority on environmental cancer, has suggested that congenital cancers and cancers in infants may be related to the action of cancer-producing agents to which the mother has been exposed during pregnancy and which penetrate the placenta to act on the rapidly developing foetal tissues. Experiments show that the younger the animal is when it is subjected to a cancer-producing agent the more certain is the production of cancer. Dr Francis Ray of the University of Florida has warned that

we may be initiating cancer in the children of today by the addition of chemicals [to food] ... We will not know, perhaps for a generation or two, what the effects will be.

The problem that concerns us here is whether any of the

chemicals we are using in our attempts to control nature play a direct or indirect role as causes of cancer. In terms of evidence gained from animal experiments we shall see that five or possibly six of the pesticides must definitely be rated as carcinogens. The list is greatly lengthened if we add those considered by some physicians to cause leukaemia in human patients. Here the evidence is circumstantial, as it must be since we do not experiment on human beings, but it is nonetheless impressive. Still other pesticides will be added as we include those whose action on living tissues or cells may be considered an indirect cause of malignancy.

One of the earliest pesticides associated with cancer is arsenic, occurring in sodium arsenite as a weed killer, and in calcium arsenate and various other compounds as insecticides. The association between arsenic and cancer in man and animals is historic. A fascinating example of the consequences of exposure to arsenic is related by Dr Hueper in his *Occupational Tumors*, a classic monograph on the subject. The city of Reichenstein in Silesia had been for almost a thousand years the site of mining for gold and silver ores, and for several hundred years for arsenic ores. Over the centuries arsenic wastes accumulated in the vicinity of the mine shafts and were picked up by streams coming down from the mountains. The underground water also became contaminated, and arsenic entered the drinking water. For centuries many of the inhabitants of this region suffered from what came to be known as 'the Reichenstein disease' – chronic arsenicism with accompanying disorders of the liver, skin, and gastro-intestinal and nervous systems. Malignant tumours were a common accompaniment of the disease. Reichenstein's disease is now chiefly of historic interest, for new water supplies were provided a quarter of a century ago, from which arsenic was largely eliminated. In Córdoba Province in Argentina, however, chronic arsenic poisoning, accompanied by arsenical skin cancers, is endemic because of the contamination of drinking water derived from rock formations containing arsenic.

It would not be difficult to create conditions similar to those in Reichenstein and Córdoba by long-continued use of arsenical insecticides. In the United States the arsenic-drenched soils of tobacco plantations, of many orchards in the North-west, and of blueberry lands in the East may easily lead to pollution of water supplies.

An arsenic-contaminated environment affects not only man but animals as well. A report of great interest came from Germany in 1936. In the area about Freiberg, Saxony, smelters for silver and lead poured arsenic fumes into the air, to drift out over the surrounding countryside and settle down upon the vegetation. According to Dr Hueper, horses, cows, goats, and pigs, which of course fed on this vegetation, showed loss of hair and thickening of the skin. Deer inhabiting nearby forests sometimes had abnormal pigment spots and precancerous warts. One had a definitely cancerous lesion. Both domestic and wild animals were affected by 'arsenical enteritis, gastric ulcers, and cirrhosis of the liver'. Sheep kept near the smelters developed cancers of the nasal sinus; at their death arsenic was found in the brain, liver, and tumours. In the area there was also

an extraordinary mortality among insects, especially bees. After rainfalls which washed the arsenical dust from the leaves and carried it along into the water of brooks and pools, a great many fish died.

An example of a carcinogen belonging to the group of new, organic pesticides is a chemical widely used against mites and ticks. Its history provides abundant proof that, despite the supposed safeguards provided by legislation, the public can be exposed to a known carcinogen for several years before the slowly moving legal processes can bring the situation under control. The story is interesting from another standpoint, proving that what the public is asked to accept as 'safe' today may turn out tomorrow to be extremely dangerous.

When this chemical was introduced in 1955, the manufacturer applied for a tolerance which would sanction the presence of small residues on any crops that might be sprayed. As required by law, he had tested the chemical on laboratory animals and submitted the results with his application. However, scientists of the Food and Drug Administration interpreted the tests as showing a possible cancer-producing tendency and the Commissioner accordingly recommended a 'zero tolerance', which is a way of saying that no residues could legally occur on food shipped across state lines. But the manufacturer had the legal right to appeal and the case was accordingly reviewed by a committee. The committee's decision was a compromise: a tolerance of 1 part per million was to be established and the product marketed for two years, during which time further laboratory tests were to determine whether the chemical was actually a carcinogen.

Although the committee did not say so, its decision meant that the public was to act as guinea-pigs, testing the suspected carcinogen along with the laboratory dogs and rats. But laboratory animals give more prompt results, and after the two years it was evident that this miticide was indeed a carcinogen. Even at that point, in 1957, the Food and Drug Administration could not instantly rescind the tolerance which allowed residues of a known carcinogen to contaminate food consumed by the public. Another year was required for various legal procedures. Finally, in December 1958 the zero tolerance which the Commissioner had recommended in 1955 became effective.

These are by no means the only known carcinogens among pesticides. In laboratory tests on animal subjects, DDT has produced suspicious liver tumours. Scientists of the Food and Drug Administration who reported the discovery of these tumours were uncertain how to classify them, but felt there was some 'justification for considering them low-grade hepatic cell carcinomas'. Dr Hueper now gives DDT the definite rating of a 'chemical carcinogen'.

Two herbicides belonging to the carbamate group, IPC and CIPC, have been found to play a role in producing skin tumours in mice. Some of the tumours were malignant. These chemicals seem to initiate the malignant change, which may then be completed by other chemicals of types prevalent in the environment.

The weed-killer aminotriazole has caused thyroid cancer in test animals. This chemical was misused by a number of cranberry growers in 1959, producing residues on some of the marketed berries. In the controversy that followed seizure of contaminated cranberries by the Food and Drug Administration, the fact that the chemical actually is cancer producing was widely challenged, even by many medical men. The scientific facts released by the Food and Drug Administration clearly indicated the carcinogenic nature of aminotriazole in laboratory rats. When these animals were fed this chemical at the rate of 100 parts per million in the drinking water (or one teaspoonful of chemical in ten thousand teaspoonfuls of water) they began to develop thyroid tumours at the 68th week. After two years, such tumours were present in more than half the rats examined. They were diagnosed as various types of benign and malignant growths. The tumours also appeared at lower levels of feeding – in fact, *a level that produced no effect was not found.* No one knows, of course, the level at which aminotriazole may be carcinogenic for man, but as a professor of medicine at Harvard University, Dr David Rutstein, has pointed out, the level is just as likely to be to man's disfavour as to his advantage.

As yet insufficient time has elapsed to reveal the full effect of the new chlorinated hydrocarbon insecticides and of the modern herbicides. Most malignancies develop so slowly that they may require a considerable segment of the victim's life to reach the stage of showing clinical symptoms. In the early 1920s women who painted luminous figures on watch dials swallowed minute amounts of radium by touching the brushes to their lips; in some of these women bone cancers developed after a lapse of 15 or more years. A period of 15 to 30 years or even more has been demonstrated for some cancers caused by occupational exposures to chemical carcinogens.

In contrast to these industrial exposures to various carcinogens the first exposures to DDT date from about 1942 for military personnel and from about 1954 for civilians, and it was not until the early fifties that a wide variety of pesticidal chemicals came into use. The full maturing of whatever seeds of malignancy have been sown by these chemicals is yet to come.

There is, however, one presently known exception to the fact that a long period of latency is common to most malignancies. This exception is leukaemia. Survivors of Hiroshima began to develop leukaemia only three years after the atomic bombing, and there is now reason to believe the latent period may be considerably shorter. Other types of cancer may in time be found to have a relatively short latent period, also, but at present leukaemia seems to be the exception to the general rule of extremely slow development.

Within the period covered by the rise of modern pesticides,

the incidence of leukaemia has been steadily rising. Figures available from the National Office of Vital Statistics clearly establish a disturbing rise in malignant diseases of the blood-forming tissues. In the year 1960, leukaemia alone claimed 12,290 victims. Deaths from all types of malignancies of blood and lymph totalled 25,400, increasing sharply from the 16,690 figure of 1950. In terms of deaths per 100,000 of population, the increase is from 11·1 in 1950 to 14·1 in 1960. The increase is by no means confined to the United States; in all countries the recorded deaths from leukaemia at all ages are rising at a rate of 4 to 5 per cent a year. What does it mean? To what lethal agent or agents, new to our environment, are people now exposed with increasing frequency?

Such world-famous institutions as the Mayo Clinic admit hundreds of victims of these diseases of the blood-forming organs. Dr Malcolm Hargraves and his associates in the Hematology Department at the Mayo Clinic report that almost without exception these patients have had a history of exposure to various toxic chemicals, including sprays which contain DDT, chlordane, benzene, lindane, and petroleum distillates.

Environmental diseases related to the use of various toxic substances have been increasing, 'particularly during the past ten years', Dr Hargraves believes. From extensive clinical experience he believes that

the vast majority of patients suffering from the blood dyscrasias and lymphoid diseases have a significant history of exposure to the various hydrocarbons, which in turn includes most of the pesticides of today. A careful medical history will almost invariably establish such a relationship.

This specialist now has a large number of detailed case histories based on every patient he has seen with leukaemias, aplastic anaemias, Hodgkin's disease, and other disorders of the blood and blood-forming tissues. 'They had all been exposed to these environmental agents, with a fair amount of exposure,' he reports.

What do these case histories show? One concerned a housewife who abhorred spiders. In mid-August she had gone into her basement with an aerosol spray containing DDT and petroleum distillate. She sprayed the entire basement thoroughly, under the stairs, in the fruit cupboards and in all the protected areas around ceiling and rafters. As she finished the spraying she began to feel quite ill, with nausea and extreme anxiety and nervousness. Within the next few days she felt better, however, and apparently not suspecting the cause of her difficulty, she repeated the entire procedure in September, running through two more cycles of spraying, falling ill, recovering temporarily, spraying again. After the third use of the aerosol new symptoms developed: fever, pains in the joints and general malaise, acute phlebitis in one leg. When examined by Dr Hargraves she was found to be suffering from acute leukaemia. She died within the following month.

Another of Dr Hargraves's patients was a professional man who had his office in an old building infested by roaches. Becoming embarrassed by the presence of these insects, he took control measures in his own hands. He spent most of one Sunday spraying the basement and all secluded areas. The spray was a 25 per cent DDT concentrate suspended in a solvent containing methylated naphthalenes. Within a short time he began to bruise and bleed. He entered the clinic bleeding from a number of haemorrhages. Studies of his blood revealed a severe depression of the bone marrow called aplastic anaemia. During the next five and one half months he received 59 transfusions in addition to other therapy. There was partial recovery but about nine years later a fatal leukaemia developed.

Where pesticides are involved, the chemicals that figure most prominently in the case histories are DDT, lindane, benzene hexachloride, the nitrophenols, the common moth crystal paradichlorobenzene, chlordane, and, of course, the solvents in which they are carried. As this physician emphasizes, pure exposure to a single chemical is the exception, rather than the rule. The commercial product usually contains combinations of several chemicals, suspended in a petroleum distillate plus some dispensing agent. The aromatic cyclic and unsaturated hydrocarbons of the vehicle may themselves be a major factor in the damage done to the blood-forming organs. From the practical rather than the medical standpoint this distinction is of little importance, however, because these petroleum solvents are an inseparable part of most common spraying practices.

The medical literature of this and other countries contains many significant cases that support Dr Hargraves's belief in a cause-and-effect relation between these chemicals and leukaemia and other blood disorders. They concern such everyday people as farmers caught in the 'fallout' of their own spray rigs or of planes, a college student who sprayed his study for ants and remained in the room to study, a woman who had installed a portable lindane vaporizer in her home, a worker in a cotton field that had been sprayed with chlordane and toxaphene. They carry, half concealed within their medical terminology, stories of such human tragedies as that of two young cousins in Czechoslovakia, boys who lived in the same town and had always worked and played together. Their last and most fateful employment was at a farm cooperative where it was their job to unload sacks of an insecticide (benzene hexachloride). Eight months later one of the boys was stricken with acute leukaemia. In nine days he was dead. At about this time his cousin began to tire easily and to run a temperature. Within about three months his symptoms became more severe and he, too, was hospitalized. Again the diagnosis was acute leukaemia, and again the disease ran its inevitably fatal course.

And then there is the case of a Swedish farmer, strangely reminiscent of that of the Japanese fisherman Kuboyama of the tuna vessel the *Lucky Dragon*. Like Kuboyama, the farmer had been a healthy man, gleaning his living from the land as Kuboyama had taken his from the sea. For each man a poison drifting out of the sky carried a death sentence. For one, it was radiation-poisoned ash; for the other, chemical dust. The farmer had treated about 60 acres of land with a dust containing DDT and benzene hexachloride. As he worked puffs of wind brought little clouds of dust swirling about him.

In the evening he felt unusually tired, and during the subsequent days he had a general feeling of weakness, with backache and aching legs as well as chills, and was obliged to take to his bed [says a report from the Medical Clinic at Lund]. His condition became worse, however, and on May 19 [a week after the spraying] he applied for admission to the local hospital.

He had a high fever and his blood count was abnormal. He was transferred to the Medical Clinic, where, after an illness of two and one half months, he died. A post mortem examination revealed a complete wasting away of the bone marrow.

How a normal and necessary process such as cell division can become altered so that it is alien and destructive is a problem that has engaged the attention of countless scientists and untold sums of money. What happens in a cell to change its orderly multiplication into the wild and uncontrolled proliferation of cancer?

When answers are found they will almost certainly be multiple. Just as cancer itself is a disease that wears many guises, appearing in various forms that differ in their origin, in the course of their development, and in the factors that influence their growth or regression, so there must be a corresponding variety of causes. Yet underlying them all, perhaps, only a few basic kinds of injuries to the cell are responsible. Here and there, in research widely scattered and sometimes not undertaken as a cancer study at all, we see glimmerings of the first light that may one day illuminate this problem.

Again we find that only by looking at some of the smallest units of life, the cell and its chromosomes, can we find that wider vision needed to penetrate such mysteries. Here, in this microcosm, we must look for those factors that somehow shift the marvellously functioning mechanisms of the cell out of their normal patterns.

One of the most impressive theories of the origin of cancer cells was developed by a German biochemist, Professor Otto Warburg of the Max Planck Institute of Cell Physiology. Warburg has devoted a lifetime of study to the complex processes of oxidation within the cell. Out of this broad background of understanding came a fascinating and lucid explanation of the way a normal cell can become malignant.

Warburg believes that either radiation or a chemical carcinogen acts by destroying the respiration of normal cells, thus depriving them of energy. This action may result from minute doses often repeated. The effect, once achieved, is irreversible. The cells not killed outright by the impact of such a respiratory poison struggle to compensate for the loss of energy. They can no longer carry on that extraordinary and efficient cycle in which vast amounts of ATP are produced, but are thrown back on a primitive and far less efficient method, that of fermentation. The struggle to survive by fermentation continues for a long period of time. It continues through ensuing cell divisions, so that all the descendant cells have this abnormal method of respiration. Once a cell has lost its normal respiration it cannot regain it – not in a year, not in a decade or in many decades. But little by little, in this gruelling struggle to restore lost energy, those cells that survive begin to compensate by increased fermentation. It is a Darwinian struggle, in which only the most fit or adaptable survive. At last they reach the point where fermentation is able to produce as much energy as respiration. At this point, cancer cells may be said to have been created from normal body cells.

Warburg's theory explains many otherwise puzzling things. The long latent period of most cancers is the time required for the infinite number of cell divisions during which fermentation is gradually increasing after the initial damage to respiration. The time required for fermentation to become dominant varies in different species because of different fermentation rates: a short time in the rat, in which cancers appear quickly, a long time (decades even) in man, in whom the development of malignancy is a deliberate process.

The Warburg theory also explains why repeated small doses of a carcinogen are more dangerous under some circumstances than a single large dose. The latter may kill the cells outright, whereas the small doses allow some to survive, though in a damaged condition. These survivors may then develop into cancer cells. This is why there is no 'safe' dose of a carcinogen.

In Warburg's theory we also find explanation of an otherwise incomprehensible fact – that one and the same agent can be useful in treating cancer and can also cause it. This, as everyone knows, is true of radiation, which kills cancer cells but may also cause cancer. It is also true of many of the chemicals now used against cancer. Why? Both types of agents damage respiration. Cancer cells already have a defective respiration, so with additional damage they die. The normal cells, suffering respiratory damage for the first time, are not killed but are set on the path that may eventually lead to malignancy.

Warburg's ideas received confirmation in 1953 when other

workers were able to turn normal cells into cancer cells merely by depriving them of oxygen intermittently over long periods. Then in 1961 other confirmation came, this time from living animals rather than tissue cultures. Radioactive tracer substances were injected into cancerous mice. Then by careful measurements of their respiration, it was found that the fermentation rate was markedly above normal, just as Warburg had foreseen.

Measured by the standards established by Warburg, most pesticides meet the criterion of the perfect carcinogen too well for comfort. As we have seen in the preceding chapter, many of the chlorinated hydrocarbons, the phenols, and some herbicides interfere with oxidation and energy production within the cell. By these means they may be creating sleeping cancer cells, cells in which an irreversible malignancy will slumber long and undetected until finally – its cause long forgotten and even unsuspected – it flares into the open as recognizable cancer.

Another path to cancer may be by way of the chromosomes. Many of the most distinguished research men in this field look with suspicion on any agent that damages the chromosomes, interferes with cell division, or causes mutations. In the view of these men any mutation is a potential cause of cancer. Although discussions of mutations usually refer to those in the germ cells, which may then make their effect felt in future generations, there may also be mutations in the body cells. According to the mutation theory of the origin of cancer, a cell, perhaps under the influence of radiation or of a chemical, develops a mutation that allows it to escape the controls the body normally asserts over cell division. It is therefore able to multiply in a wild and unregulated manner. The new cells resulting from these divisions have the same ability to escape control, and in time enough such cells have accumulated to constitute a cancer.

Other investigators point to the fact that the chromosomes in cancer tissue are unstable; they tend to be broken or damaged, the number may be erratic, there may even be double sets.

The first investigators to trace chromosome abnormalities all the way to actual malignancy were Albert Levan and John J. Biesele, working at the Sloan-Kettering Institute in New York. As to which came first, the malignancy or the disturbance of the chromosomes, these workers say without hesitation that 'the chromosomal irregularities precede the malignancy'. Perhaps, they speculate, after the initial chromosome damage and the resulting instability there is a long period of trial and error through many cell generations (the long latent period of malignancy) during which a collection of mutations is finally accumulated which allow the cells to escape from control and embark on the unregulated multiplication that is cancer.

Ojvind Winge, one of the early proponents of the theory of chromosome instability, felt that chromosome doublings were especially significant. Is it coincidence, then, that benzene hexachloride and its relative, lindane, are known through repeated observations to double the chromosomes in experimental plants – and that these same chemicals have been implicated in many well-documented cases of fatal anaemias? And what of the many other pesticides that interfere with cell division, break chromosomes, cause mutations?

It is easy to see why leukaemia should be one of the most common diseases to result from exposure to radiation or to chemicals that imitate radiation. The principal targets of physical or chemical mutagenic agents are cells that are undergoing especially active division. This includes various tissues but most importantly those engaged in the production of blood. The bone marrow is the chief producer of red blood cells throughout life, sending some ten million new cells per second into the bloodstream of man. White corpuscles are formed in the lymph glands and in some of the marrow cells at a variable, but still prodigious, rate.

Certain chemicals, again reminding us of radiation products like Strontium 90, have a peculiar affinity for the bone marrow. Benzene, a frequent constituent of insecticidal solvents, lodges in the marrow and remains deposited there for periods known to be as long as 20 months. Benzene itself has been recognized in medical literature for many years as a cause of leukaemia.

The rapidly growing tissues of a child would also afford conditions most suitable for the development of malignant cells. Sir Macfarlane Burnet has pointed out that not only is leukaemia increasing throughout the world but it has become most common in the three- to four-year age bracket, an age incidence shown by no other disease. According to this authority,

The peak between three and four years of age can hardly have any other interpretation than exposure of the young organism to a mutagenic stimulus around the time of birth.

Another mutagen known to produce cancer is urethane. When pregnant mice are treated with this chemical not only do they develop cancer of the lung but their young do, also. The only exposure of the infant mice to urethane was pre-natal in these experiments, proving that the chemical must have passed through the placenta. In human populations exposed to urethane or related chemicals there is a possibility that tumours will develop in infants through pre-natal exposure, as Dr Hueper has warned.

Urethane as a carbamate is chemically related to the herbicides IPC and CIPC. Despite the warnings of cancer experts, carbamates are now widely used, not only as insecticides, weed killers, and fungicides, but also in a variety of products including plasticizers, medicines, clothing, and insulating materials.

The road to cancer may also be an indirect one. A substance that is not a carcinogen in the ordinary sense may disturb the normal functioning of some part of the body in such a way that malignancy results. Important examples are the cancers, especially of the reproductive system, that appear to be linked with disturbances of the balance of sex hormones; these disturbances, in turn, may in some cases be the result of something that affects the ability of the liver to preserve a proper level of these hormones. The chlorinated hydrocarbons are precisely the kind of agent that can bring about this kind of indirect carcinogenesis, because all of them are toxic in some degree to the liver.

The sex hormones are, of course, normally present in the body and perform a necessary growth-stimulating function in relation to the various organs of reproduction. But the body has a built-in protection against excessive accumulations, for the liver acts to keep a proper balance between male and female hormones (both are produced in the bodies of both sexes, although in different amounts) and to prevent an excess accumulation of either. It cannot do so, however, if it has been damaged by disease or chemicals, or if the supply of the B-complex vitamins has been reduced. Under these conditions the oestrogens build up to abnormally high levels.

What are the effects? In animals, at least, there is abundant evidence from experiments. In one such, an investigator at the Rockefeller Institute for Medical Research found that rabbits with livers damaged by disease show a very high incidence of uterine tumours, thought to have developed because the liver was no longer able to inactivate the oestrogens in the blood, so that they 'subsequently rose to a carcinogenic level'. Extensive experiments on mice, rats, guinea-pigs, and monkeys show that prolonged administration of oestrogens (not necessarily at high levels) has caused changes in the tissues of the reproductive organs, 'varying from benign overgrowths to definite malignancy'. Tumours of the kidneys have been induced in hamsters by administering oestrogens.

Although medical opinion is divided on the question, much evidence exists to support the view that similar effects may occur in human tissues. Investigators at the Royal Victoria Hospital at McGill University found two-thirds of 150 cases of uterine cancer studied by them gave evidence of abnormally high oestrogen levels. In 90 per cent of a later series of 20 cases there was similar high oestrogen activity.

It is possible to have liver damage sufficient to interfere with oestrogen elimination without detection of the damage by any tests now available to the medical profession. This can easily be caused by the chlorinated hydrocarbons, which, as we have seen, set up changes in liver cells at very low levels of intake. They also cause loss of the B vitamins. This, too, is extremely important, for other chains of evidence show the protective role of these vitamins against cancer. The late C. P. Rhoads, one-time director of the Sloan-Kettering Institute for Cancer Research, found that test animals exposed to a very potent chemical carcinogen developed no cancer if they had been fed yeast, a rich source of the natural B vitamins. A deficiency of these vitamins has been found to accompany mouth cancer and perhaps cancer of other sites in the digestive tract. This has been observed not only in the United States but in the far northern parts of Sweden and Finland, where the diet is ordinarily deficient in vitamins. Groups prone to primary liver cancer, as for example the Bantu tribes of Africa, are typically subject to malnutrition. Cancer of the male breast is also prevalent in parts of Africa, associated with liver disease and malnutrition. In post-war Greece enlargement of the male breast was a common accompaniment of periods of starvation.

In brief, the argument for the indirect role of pesticides in cancer is based on their proven ability to damage the liver and to reduce the supply of B vitamins, thus leading to an increase in the 'endogenous' oestrogens, or those produced by the body itself. Added to these are the wide variety of synthetic oestrogens to which we are increasingly exposed – those in cosmetics, drugs, foods, and occupational exposures. The combined effect is a matter that warrants the most serious concern.

Human exposures to cancer-producing chemicals (including pesticides) are uncontrolled and they are multiple. An individual may have many different exposures to the same chemical. Arsenic is an example. It exists in the environment of every individual in many different guises: as an air pollutant, a contaminant of water, a pesticide residue on food, in medicines, cosmetics, wood preservatives, or as a colouring agent in paints and inks. It is quite possible that no one of these exposures alone would be sufficient to precipitate malignancy – yet any single supposedly 'safe dose' may be enough to tip the scales that are already loaded with other 'safe doses'.

Or again the harm may be done by two or more different carcinogens acting together, so that there is a summation of their effects. The individual exposed to DDT, for example, is almost certain to be exposed to other liver-damaging hydrocarbons, which are so widely used as solvents, paint removers, degreasing agents, dry-cleaning fluids, and anaesthetics. What then can be a 'safe dose' of DDT?

The situation is made even more complicated by the fact that one chemical may act on another to alter its effect. Cancer may sometimes require the complementary action of two chemicals, one of which sensitizes the cell or tissue so that it may later, under the action of another or promoting agent, develop true malignancy. Thus, the herbicides IPC and CIPC may act as initiators in the production of skin tumours, sowing the seeds of malignancy that may be brought into actual being by something else – perhaps a common detergent.

There may be interaction, too, between a physical and a chemical agent. Leukaemia may occur as a two-step process, the malignant change being initiated by X-radiation, the promoting action being supplied by a chemical, as, for example, urethane. The growing exposure of the population to radiation from various sources plus the many contacts with a host of chemicals suggest a grave new problem for the modern world.

The pollution of water supplies with radioactive materials poses another problem. Such materials, present as contaminants in water that also contains chemicals, may actually change the nature of the chemicals by the impact of ionizing radiation, re-arranging their atoms in unpredictable ways to create new chemicals.

Water-pollution experts throughout the United States are concerned by the fact that detergents are now a troublesome and practically universal contaminant of public water supplies. There is no practical way to remove them by treatment. Few detergents are known to be carcinogenic, but in an indirect way they may promote cancer by acting on the lining of the digestive tract, changing the tissues so that they more easily absorb dangerous chemicals, thereby aggravating their effect. But who can foresee and control this action? In the kaleidoscope of shifting conditions, what dose of a carcinogen can be 'safe' except a zero dose?

We tolerate cancer-causing agents in our environment at our peril, as was clearly illustrated by a recent happening. In the spring of 1961 an epidemic of liver cancer appeared among rain-bow trout in many federal, state, and private hatcheries. Trout in both eastern and western parts of the United States were affected; in some areas practically 100 per cent of the trout over three years of age developed cancer. This discovery was made because of a pre-existing arrangement between the Environmental Cancer Section of the National Cancer Institute and the Fish and Wildlife Service for the reporting of all fish with tumours, so that early warning might be had of a cancer hazard to man from water contaminants.

Although studies are still under way to determine the exact cause of this epidemic over so wide an area, the best evidence is said to point to some agent present in the prepared hatchery feeds. These contain an incredible variety of chemical additives and medicinal agents in addition to the basic foodstuffs.

The story of the trout is important for many reasons, but chiefly as an example of what can happen when a potent carcinogen is introduced into the environment of any species. Dr Hueper has described this epidemic as a serious warning that greatly increased attention must be given to controlling the number and variety of environmental carcinogens. 'If such preventive measures are not taken,' says Dr Hueper, 'the stage will be set at a progressive rate for the future occurrence of a similar disaster to the human population.'

The discovery that we are, as one investigator phrased it, living in a 'sea of carcinogens' is of course dismaying and may easily lead to reactions of despair and defeatism. 'Isn't it a hopeless situation?' is the common reaction. 'Isn't it impossible even to attempt to eliminate these cancer-producing agents from our world? Wouldn't it be better not to waste time trying, but instead to put all our efforts into research to find a cure for cancer?'

When this question is put to Dr Hueper, whose years of distinguished work in cancer make his opinion one to respect, his reply is given with the thoughtfulness of one who has pondered it long, and has a lifetime of research and experience behind his judgement. Dr Hueper believes that our situation with regard to cancer today is very similar to that which faced mankind with regard to infectious diseases in the closing years of the nineteenth century. The causative relation between pathogenic organisms and many diseases had been established through the brilliant work of Pasteur and Koch. Medical men and even the general public were becoming aware that the human environment was inhabited by an enormous number of micro-organisms capable of causing disease, just as today carcinogens pervade our surroundings. Most infectious diseases have now been brought under a reasonable degree of control and some have been practically eliminated. This brilliant medical achievement came about by an attack that was twofold – that stressed prevention as well as cure. Despite the prominence that 'magic bullets' and 'wonder drugs' hold in the layman's mind, most of the really decisive battles in the war against infectious disease consisted of measures to eliminate disease organisms from the environment. An example from history concerns the great outbreak of cholera in London more than one hundred years ago. A London physician, John Snow, mapped the occurrence of cases and found they originated in one area, all of whose inhabitants drew their water from one pump located in Broad Street. In a swift and decisive practice of preventive medicine, Dr Snow removed the handle from the pump. The epidemic was thereby brought under control – not by a magic pill that killed the (then unknown) organism of cholera, but by eliminating the organism from the environment. Even therapeutic measures have the important result not only of curing the patient but of reducing the foci of infection. The present comparative rarity of tuberculosis results in large measure from the fact that the average person now seldom comes into contact with the tubercle bacillus.

Today we find our world filled with cancer-producing agents. An attack on cancer that is concentrated wholly or even largely on therapeutic measures (even assuming a 'cure' could be found) in Dr Hueper's opinion will fail because it leaves untouched the great reservoirs of carcinogenic agents which would continue to claim new victims faster than the as yet elusive 'cure' could allay the disease.

Why have we been slow to adopt this common-sense approach to the cancer problem? Probably 'the goal of curing the victims of cancer is more exciting, more tangible, more glamorous and rewarding than prevention,' says Dr Hueper. Yet to prevent cancer from ever being formed is 'definitely more humane' and can be 'much more effective than cancer cures'. Dr Hueper has little patience with the wishful thinking that promises 'a magic pill that we shall take each morning before breakfast' as protection against cancer. Part of the public trust in such an eventual outcome results from the misconception that cancer is a single, though mysterious disease, with a single cause and, hopefully, a single cure. This of course is far from the known truth. Just as environmental cancers are induced by a wide variety of chemical and physical agents, so the malignant condition itself is manifested in many different and biologically distinct ways.

The long-promised 'breakthrough', when or if it comes, cannot be expected to be a panacea for all types of malignancy.

Although the search must be continued for therapeutic measures to relieve and to cure those who have already become victims of cancer, it is a disservice to humanity to hold out the hope that the solution will come suddenly, in a single master stroke. It will come slowly, one step at a time. Meanwhile as we pour our millions into research and invest all our hopes in vast programmes to find cures for established cases of cancer, we are neglecting the golden opportunity to prevent, even while we seek to cure.

The task is by no means a hopeless one. In one important respect the outlook is more encouraging than the situation regarding infectious disease at the turn of the century. The world was then full of disease germs, as today it is full of carcinogens. But man did not put the germs into the environment and his role in spreading them was involuntary. In contrast, man *has* put the vast majority of carcinogens into the environment, and he can, if he wishes, eliminate many of them. The chemical agents of cancer have become entrenched in our world in two ways: first, and ironically, through man's search for a better and easier way of life; second, because the manufacture and sale of such chemicals has become an accepted part of our economy and our way of life.

It would be unrealistic to suppose that all chemical carcinogens can or will be eliminated from the modern world. But a very large proportion are by no means necessities of life. By their elimination the total load of carcinogens would be enormously lightened, and the threat that one in every four will develop cancer would at least be greatly mitigated. The most determined effort should be made to eliminate those carcinogens that now contaminate our food, our water supplies, and our atmosphere, because these provide the most dangerous type of contact – minute exposures, repeated over and over throughout the years.

Among the most eminent men in cancer research are many others who share Dr Hueper's belief that malignant diseases can be reduced significantly by determined efforts to identify the environmental causes and to eliminate them or reduce their impact. For those in whom cancer is already a hidden or a visible presence, efforts to find cures must of course continue. But for those not yet touched by the disease and certainly for the generations as yet unborn, prevention is the imperative need.

Chapter 15

Nature Fights Back

To have risked so much in our efforts to mould nature to our satisfaction and yet to have failed in achieving our goal would indeed be the final irony. Yet this, it seems, is our situation. The truth, seldom mentioned but there for anyone to see, is that nature is not so easily moulded and that the insects are finding ways to circumvent our chemical attacks on them.

The insect world is nature's most astonishing phenomenon [said the Dutch biologist C. J. Briejèr]. Nothing is impossible to it; the most improbable things commonly occur there. One who penetrates deeply into its mysteries is continually breathless with wonder. He knows that anything can happen, and that the completely impossible often does.

The 'impossible' is now happening on two broad fronts. By a process of genetic selection, the insects are developing strains resistant to chemicals. This will be discussed in the following chapter. But the broader problem, which we shall look at now, is the fact that our chemical attack is weakening the defences inherent in the environment itself, defences designed to keep the various species in check. Each time we breach these defences a horde of insects pours through.

From all over the world come reports that make it clear we are in a serious predicament. At the end of a decade or more of intensive chemical control, entomologists were finding that problems they had considered solved a few years earlier had returned to plague them. And new problems had arisen as insects once present only in insignificant numbers had increased to the status of serious pests. By their very nature chemical controls are self-defeating, for they have been devised and applied without taking into account the complex biological systems against which they have been blindly hurled. The chemicals may have been pre-tested against a few individual species, but not against living communities.

In some quarters nowadays it is fashionable to dismiss the balance of nature as a state of affairs that prevailed in an earlier, simpler world – a state that has now been so thoroughly upset that we might as well forget it. Some find this a convenient assumption, but as a chart for a course of action it is highly dangerous. The balance of nature is not the same today as in Pleistocene times, but it is still there: a complex, precise, and highly integrated system of relationships between living things which cannot safely be ignored any more than the law of gravity can be defied with impunity by a man perched on the edge of a cliff. The balance of nature is not a *status quo*; it is fluid, ever shifting, in a constant state of adjustment. Man, too, is part of this balance. Sometimes the balance is in his favour; sometimes – and all too often through his own activities – it is shifted to his disadvantage.

Two critically important facts have been overlooked in designing the modern insect-control programmes. The first is that the really effective control of insects is that applied by nature, not by man. Populations are kept in check by something the ecologists call the resistance of the environment, and this has been so since the first life was created. The amount of food available, conditions of weather and climate, the presence of competing or predatory species, all are critically important. 'The greatest single factor in preventing insects from overwhelming the rest of the world is the internecine warfare which they carry out among themselves,' said the entomologist Robert Metcalf. Yet most of the chemicals now used kill all insects, our friends and enemies alike.

The second neglected fact is the truly explosive power of a species to reproduce once the resistance of the environment has been weakened. The fecundity of many forms of life is almost beyond our power to imagine, though now and then we have suggestive glimpses. I remember from student days the miracle that could be wrought in a jar containing a simple mixture of hay and water merely by adding to it a few drops of material from a mature culture of protozoa. Within a few days the jar would contain a whole galaxy of whirling, darting life –

uncountable trillions of the slipper animalcule, *Paramecium*, each small as a dust grain, all multiplying without restraint in their temporary Eden of favourable temperatures, abundant food, absence of enemies. Or I think of shore rocks white with barnacles as far as the eye can see, or of the spectacle of passing through an immense school of jellyfish, mile after mile, with seemingly no end to the pulsing, ghostly forms scarcely more substantial than the water itself.

We see the miracle of nature's control at work when the cod move through winter seas to their spawning grounds, where each female deposits several millions of eggs. The sea does not become a solid mass of cod as it would surely do if all the progeny of all the cod were to survive. The checks that exist in nature are such that out of the millions of young produced by each pair only enough, on the average, survive to adulthood to replace the parent fish.

Biologists used to entertain themselves by speculating as to what would happen if, through some unthinkable castastrophe, the natural restraints were thrown off and all the progeny of a single individual survived. Thus Thomas Huxley a century ago calculated that a single female aphis (which has the curious power of reproducing without mating) could produce progeny in a single year's time whose total weight would equal that of the inhabitants of the Chinese Empire of his day.

Fortunately for us such an extreme situation is only theoretical, but the dire results of upsetting nature's own arrangements are well known to students of animal populations. The stockman's zeal for eliminating the coyote has resulted in plagues of field mice, which the coyote formerly controlled. The oft-repeated story of the Kaibab deer in Arizona is another case in point. At one time the deer population was in equilibrium with its environment. A number of predators – wolves, pumas, and coyotes – prevented the deer from outrunning their food supply. Then a campaign was begun to 'conserve' the deer by killing off their enemies. Once the predators were gone, the deer increased prodigiously and soon there was not enough food for them. The browse line on the trees went higher and higher as they sought food, and in time many more deer were dying of starvation than had formerly been killed by predators. The whole environment, moreover, was damaged by their desperate efforts to find food.

The predatory insects of field and forests play the same role as the wolves and coyotes of the Kaibab. Kill them off and the population of the prey insect surges upwards.

No one knows how many species of insects inhabit the earth because so many are yet to be identified. But more than 700,000 have already been described. This means that in terms of the number of species, 70 to 80 per cent of the earth's creatures are insects. The vast majority of these insects are held in check by natural forces, without any intervention by man. If this were not so, it is doubtful that any conceivable volume of chemicals – or any other methods – could possibly keep down their populations.

The trouble is that we are seldom aware of the protection afforded by natural enemies until it fails. Most of us walk unseeing through the world, unaware alike of its beauties, its wonders, and the strange and sometimes terrible intensity of the lives that are being lived about us. So it is that the activities of the insect predators and parasites are known to few. Perhaps we may have noticed an oddly shaped insect of ferocious mien on a bush in the garden and been dimly aware that the praying mantis lives at the expense of other insects. But we see with understanding eye only if we have walked in the garden at night and here and there with a flashlight have glimpsed the mantis stealthily creeping upon her prey. Then we sense something of the drama of the hunter and the hunted. Then we begin to feel something of that relentlessly pressing force by which nature controls her own.

The predators – insects that kill and consume other insects – are of many kinds. Some are quick and with the speed of swallows snatch their prey from the air. Others plod methodically along a stem, plucking off and devouring sedentary insects like the aphids. The yellowjackets capture soft-bodied insects and feed the juices to their young. Muddauber wasps build columned nests of mud under the eaves of houses and stock them with insects on which their young will feed. The horseguard wasp hovers above herds of grazing cattle, destroying the bloodsucking flies that torment them. The loudly buzzing syrphid fly, often mistaken for a bee, lays its eggs on leaves of aphis-infested plants; the hatching larvae then consume immense numbers of aphids. Ladybugs or lady beetles are among the most effective destroyers of aphids, scale insects, and other plant-eating insects. Literally hundreds of aphids are consumed by a single ladybug to stoke the little fires of energy which she requires to produce even a single batch of eggs.

Even more extraordinary in their habits are the parasitic insects. These do not kill their hosts outright. Instead, by a variety of adaptations they utilize their victims for the nurture of their own young. They may deposit their eggs within the larvae or eggs of their prey, so that their own developing young may find food by consuming the host. Some attach their eggs to a caterpillar by means of a sticky solution; on hatching, the larval parasite bores through the skin of the host. Others, led by an instinct that simulates foresight, merely lay their eggs on a leaf so that a browsing caterpillar will eat them inadvertently.

Everywhere, in field and hedgerow and garden and forest, the insect predators and parasites are at work. Here, above a pond, the dragonflies dart and the sun strikes fire from their wings. So their ancestors sped through swamps where huge reptiles lived. Now, as in those ancient times, the sharp-eyed dragonflies capture mosquitoes in the air, scooping them in with basket-shaped legs. In the waters below, their young, the dragonfly nymphs, or naiads, prey on the aquatic stages of mosquitoes and other insects.

Or there, almost invisible against a leaf, is the lacewing, with green gauze wings and golden eyes, shy and secretive, descendant of an ancient race that lived in Permian times. The adult lacewing feeds mostly on plant nectars and the honeydew of aphids, and in time she lays her eggs, each on the end of a long stalk which she fastens to a leaf. From these emerge her children – strange, bristled larvae called aphis lions, which live by preying on aphids, scales, or mites, which they capture and suck dry of fluid. Each may consume several hundred aphids before the ceaseless turning of the cycle of its life brings the time when it will spin a white silken cocoon in which to pass the pupal stage.

And there are many wasps, and flies as well, whose very existence depends on the destruction of the eggs or larvae of other insects through parasitism. Some of the egg parasites are exceedingly minute wasps, yet by their numbers and their great activity they hold down the abundance of many crop-destroying species.

All these small creatures are working – working in sun and rain, during the hours of darkness, even when winter's grip has damped down the fires of life to mere embers. Then this vital force is merely smouldering, awaiting the time to flare again into activity when spring awakens the insect world. Meanwhile, under the white blanket of snow, below the frost-hardened soil, in crevices in the bark of trees, and in sheltered caves, the parasites and the predators have found ways to tide themselves over the season of cold.

The eggs of the mantis are secure in little cases of thin parchment attached to the branch of a shrub by the mother who lived her life span with the summer that is gone.

The female *Polistes* wasp, taking shelter in a forgotten corner of some attic, carries in her body the fertilized eggs, the heritage on which the whole future of her colony depends. She, the lone survivor, will start a small paper nest in the spring, lay a few eggs in its cells, and carefully rear a small force of workers. With their help she will then enlarge the nest and develop the colony. Then the workers, foraging ceaselessly through the hot days of summer, will destroy countless caterpillars.

Thus, through the circumstances of their lives, and the nature of our own wants, all these have been our allies in keeping the balance of nature tilted in our favour. Yet we have turned our artillery against our friends. The terrible danger is that we have grossly underestimated their value in keeping at bay a dark tide of enemies that, without their help, can overrun us.

The prospect of a general and permanent lowering of environmental resistance becomes grimly and increasingly real with each passing year as the number, variety, and destructiveness of insecticides grow. With the passage of time we may expect progressively more serious outbreaks of insects, both disease-carrying and crop-destroying species, in excess of anything we have ever known.

'Yes, but isn't this all theoretical?' you may ask. 'Surely it won't really happen – not in my lifetime, anyway.'

But it is happening, here and now. Scientific journals had already recorded some 50 species involved in violent dislocations of nature's balance by 1958. More examples are being found every year. A recent review of the subject contained references to 215 papers reporting or discussing unfavourable upsets in the balance of insect populations caused by pesticides.

Sometimes the result of chemical spraying has been a tremendous upsurge of the very insect the spraying was intended to control, as when blackflies in Ontario became 17 times more abundant after spraying than they had been before. Or when in England an enormous outbreak of the cabbage aphid – an outbreak that had no parallel on record – followed spraying with one of the organic phosphorus chemicals.

At other times spraying, while reasonably effective against the target insect, has let loose a whole Pandora's box of destructive pests that had never previously been abundant enough to cause trouble. The spider mite, for example, has become practically a world-wide pest as DDT and other insecticides have killed off its enemies. The spider mite is not an insect. It is a barely visible eight-legged creature belonging to the group that includes spiders, scorpions, and ticks. It has mouth parts adapted for piercing and sucking, and a prodigious appetite for the chlorophyll that makes the world green. It inserts these minute and stiletto-sharp mouth parts into the outer cells of leaves and evergreen needles and extracts the chlorophyll. A mild infestation gives trees and shrubbery a mottled or salt-and-pepper appearance; with a heavy mite population, foliage turns yellow and falls.

This is what happened in some of the western national forests a few years ago, when in 1956 the United States Forest Service sprayed some 885,000 acres of forested lands with DDT. The intention was to control the spruce budworm, but the following summer it was discovered that a problem worse than the budworm damage had been created. In surveying the forests from the air, vast blighted areas could be seen where the magnificent Douglas firs were turning brown and dropping their needles. In the Helena National Forest and on the western slopes of the Big Belt Mountains, then in other areas of Montana and down into Idaho the forests looked as though they had been scorched. It was evident that this summer of 1957 had brought the most extensive and spectacular infestation of spider mites in history.

Almost all of the sprayed area was affected. Nowhere else was the damage evident. Searching for precedents, the foresters could remember other scourges of spider mites, though less dramatic than this one. There had been similar trouble along the Madison River in Yellowstone Park in 1929, in Colorado 20 years later, and then in New Mexico in 1956. *Each of these outbreaks had followed forest spraying with insecticides.* (The 1929 spraying, occurring before the DDT era, employed lead arsenate.)

Why does the spider mite appear to thrive on insecticides? Besides the obvious fact that it is relatively insensitive to them, there seem to be two other reasons. In nature it is kept in check by various predators such as ladybugs, a gall midge, predaceous mites and several pirate bugs, all of them extremely sensitive to insecticides. The third reason has to do with population pressure within the spider mite colonies. An undisturbed colony of mites is a densely settled community, huddled under a protective webbing for concealment from its enemies. When sprayed, the colonies disperse as the mites, irritated though not killed by the chemicals, scatter out in search of places where they will not be disturbed. In so doing they find a far greater abundance of space and food than was available in the former colonies. Their enemies are now dead so there is no need for the mites to spend their energy in secreting protective webbing. Instead, they pour all their energies into producing more mites. It is not uncommon for their egg production to be increased threefold – all through the beneficent effect of insecticides.

In the Shenandoah Valley of Virginia, a famous apple-growing region, hordes of a small insect called the red-banded leaf roller arose to plague the growers as soon as DDT began to replace arsenate of lead. Its depredations had never before been important; soon its toll rose to 50 per cent of the crop and it achieved the status of the most destructive pest of apples, not only in this region but throughout much of the East and Mid-west, as the use of DDT increased.

The situation abounds in ironies. In the apple orchards of Nova Scotia in the late 1940s the worst infestations of the codling moth (cause of 'wormy apples') were in the orchards regularly sprayed. In unsprayed orchards the moths were not abundant enough to cause real trouble.

Diligence in spraying had a similarly unsatisfactory reward in the eastern Sudan, where cotton growers had a bitter experience with DDT. Some 60,000 acres of cotton were being grown under irrigation in the Gash Delta. Early trials of DDT having given apparently good results, spraying was intensified. It was then that trouble began. One of the most destructive enemies of cotton is the bollworm. But the more cotton was sprayed, the more bollworms appeared. The unsprayed cotton suffered less damage to fruits and later to mature bolls than the sprayed, and in twice-sprayed fields the yield of seed cotton dropped significantly. Although some of the leaf-feeding insects were eliminated, any benefit that might thus have been gained was more than offset by bollworm damage. In the end the growers were faced with the unpleasant truth that their cotton yield would have been greater had they saved themselves the trouble and expense of spraying.

In the Belgian Congo and Uganda the results of heavy applications of DDT against an insect pest of the coffee bush were almost 'catastrophic'. The pest itself was found to be almost completely unaffected by the DDT, while its predator was extremely sensitive.

In America, farmers have repeatedly traded one insect enemy for a worse one as spraying upsets the population dynamics of the insect world. Two of the mass-spraying programmes recently carried out have had precisely this effect. One was the fire ant eradication programme in the South; the other was the spraying for the Japanese beetle in the Mid-west. (See Chapters 10 and 7.)

When a wholesale application of heptachlor was made to the farmlands in Louisiana in 1957, the result was the unleashing of one of the worst enemies of the sugarcane crop – the sugarcane borer. Soon after the heptachlor treatment, damage by borers increased sharply. The chemical aimed at the fire ant had killed off the enemies of the borer. The crop was so severely damaged that farmers sought to bring suit against the state for negligence in not warning them that this might happen.

The same bitter lesson was learned by Illinois farmers. After the devastating bath of dieldrin recently administered to the farmlands in eastern Illinois for the control of the Japanese beetle, farmers discovered that corn borers had increased enormously in the treated area. In fact, corn grown in fields within this area contained almost twice as many of the destructive larvae of this insect as did the corn grown outside. The farmers may not yet be aware of the biological basis of what has happened, but they need no scientists to tell them they have made a poor bargain. In trying to get rid of one insect, they have brought on a scourge of a much more destructive one. According to Department of Agriculture estimates, total damage by the Japanese beetle in the United States adds up to about 10 million dollars a year, while damage by the corn borer runs to about 85 million.

It is worth noting that natural forces had been heavily relied on for control of the corn borer. Within two years after this insect was accidentally introduced from Europe in 1917, the United States Government had mounted one of its most intensive programmes for locating and importing parasites of an insect pest. Since that time 24 species of parasites of the corn borer have been brought in from Europe and the Orient at considerable expense. Of these, 5 are recognized as being of distinct value in control. Needless to say, the results of all this work are now jeopardized as the enemies of the corn borer are killed off by the sprays.

If this seems absurd, consider the situation in the citrus groves of California, where the world's most famous and successful experiment in biological control was carried out in the 1880s. In 1872 a scale insect that feeds on the sap of citrus trees appeared in California and within the next 15 years developed into a pest so destructive that the fruit crop in many orchards was a complete loss. The young citrus industry was threatened with destruction. Many farmers gave up and pulled out their trees. Then a parasite of the scale insect was imported from Australia, a small lady beetle called the vedalia. Within only two years after the first shipment of the beetles, the scale was under complete control throughout the citrus-growing sections of California. From that time on one could search for days among the orange groves without finding a single scale insect.

Then in the 1940s the citrus growers began to experiment with glamorous new chemicals against other insects. With the advent of DDT and the even more toxic chemicals to follow, the populations of the vedalia in many sections of California were wiped out. Its importation had cost the government a mere

$5,000. Its activities had saved the fruit growers several millions of dollars a year, but in a moment of heedlessness the benefit was cancelled out. Infestations of the scale insect quickly reappeared and damage exceeded anything that had been seen for fifty years.

'This possibly marked the end of an era,' said Dr Paul DeBach of the Citrus Experiment Station in Riverside. Now control of the scale has become enormously complicated. The vedalia can be maintained only by repeated releases and by the most careful attention to spray schedules, to minimize their contact with insecticides. And regardless of what the citrus growers do, they are more or less at the mercy of the owners of adjacent acreages, for severe damage has been done by insecticidal drift.

All these examples concern insects that attack agricultural crops. What of those that carry disease? There have already been warnings. On Nissan Island in the South Pacific, for example, spraying had been carried on intensively during the Second World War, but was stopped when hostilities came to an end. Soon swarms of a malaria-carrying mosquito reinvaded the island. All of its predators had been killed off and there had not been time for new populations to become established. The way was therefore clear for a tremendous population explosion. Marshall Laird, who has described this incident, compares chemical control to a treadmill; once we have set foot on it we are unable to stop for fear of the consequences.

In some parts of the world disease can be linked with spraying in quite a different way. For some reason, snail-like molluscs seem to be almost immune to the effects of insecticides. This has been observed many times. In the general holocaust that followed the spraying of salt marshes in eastern Florida (page 137), aquatic snails alone survived. The scene as described was a macabre picture – something that might have been created by a surrealist brush. The snails moved among the bodies of the dead fishes and the moribund crabs, devouring the victims of the death rain of poison.

But why is this important? It is important because many aquatic snails serve as hosts of dangerous parasitic worms that spend part of their life cycle in a mollusc, part in a human being.

Examples are the blood flukes, or schistosoma, that cause serious disease in man when they enter the body by way of drinking water or through the skin when people are bathing in infested waters. The flukes are released into the water by the host snails. Such diseases are especially prevalent in parts of Asia and Africa. Where they occur, insect-control measures that favour a vast increase of snails are likely to be followed by grave consequences.

And of course man is not alone in being subject to snail-borne disease. Liver disease in cattle, sheep, goats, deer, elk, rabbits, and various other warm-blooded animals may be caused by liver flukes that spend part of their life cycles in fresh-water snails. Livers infested with these worms are unfit for use as human food and are routinely condemned. Such rejections cost American cattlemen about $3\frac{1}{2}$ million dollars annually. Anything that acts to increase the number of snails can obviously make this problem an even more serious one.

Over the past decade these problems have cast long shadows, but we have been slow to recognize them. Most of those best fitted to develop natural controls and assist in putting them into effect have been too busy labouring in the more exciting vineyards of chemical control. It was reported in 1960 that only 2 per cent of all the economic entomologists in the country were then working in the field of biological controls. A substantial number of the remaining 98 per cent were engaged in research on chemical insecticides.

Why should this be? The major chemical companies are pouring money into the universities to support research on insecticides. This creates attractive fellowships for graduate students and attractive staff positions. Biological-control studies, on the other hand, are never so endowed – for the simple reason that they do not promise anyone the fortunes that are to be made in the chemical industry. These are left to state and federal agencies, where the salaries paid are far less.

This situation also explains the otherwise mystifying fact that certain outstanding entomologists are among the leading advocates of chemical control. Inquiry into the background of some of these men reveals that their entire research programme is supported by the chemical industry. Their professional prestige, sometimes their very jobs, depend on the perpetuation of chemical methods. Can we then expect them to bite the hand that literally feeds them? But knowing their bias, how much credence can we give to their protests that insecticides are harmless?

Amid the general acclaim for chemicals as the principal method of insect control, minority reports have occasionally been filed by those few entomologists who have not lost sight of the fact that they are neither chemists nor engineers, but biologists.

F. H. Jacob in England has declared that

the activities of many so-called economic entomologists would make it appear that they operate in the belief that salvation lies at the end of a spray nozzle ... that when they have created problems of resurgence or resistance or mammalian toxicity, the chemist will be ready with another pill. That view is not held here.... Ultimately only the biologist will provide the answers to the basic problems of pest control.

Economic entomologists must realize [wrote A. D. Pickett of Nova Scotia] that they are dealing with living things ... their work must be more than simply insecticide testing or a quest for highly destructive chemicals.

Dr Pickett himself was a pioneer in the field of working out sane methods of insect control that take full advantage of the predatory and parasitic species. The method which he and his associates evolved is today a shining model but one too little emulated. Only in the integrated control programmes developed by some California entomologists do we find anything comparable in this country.

Dr Pickett began his work some thirty-five years ago in the apple orchards of the Annapolis Valley in Nova Scotia, once one of the most concentrated fruit-growing areas in Canada. At that time it was believed that insecticides – then inorganic chemicals – would solve the problems of insect control, that the only task was to induce fruit growers to follow the recommended methods. But the rosy picture failed to materialize. Somehow the insects persisted. New chemicals were added, better spraying equipment was devised, and the zeal for spraying increased, but the insect problem did not get any better. Then DDT promised to 'obliterate the nightmare' of codling moth outbreaks. What actually resulted from its use was an unprecedented scourge of mites. 'We move from crisis to crisis, merely trading one problem for another,' said Dr Pickett.

At this point, however, Dr Pickett and his associates struck out on a new road instead of going along with other entomologists who continued to pursue the will-o'-the-wisp of the ever more toxic chemical. Recognizing that they had a strong ally in nature, they devised a programme that makes maximum use of natural controls and minimum use of insecticides. Whenever insecticides are applied only minimum dosages are used – barely enough to control the pest without avoidable harm to beneficial species. Proper timing also enters in. Thus, if nicotine sulphate is applied before rather than after the apple blossoms turn pink one of the important predators is spared, probably because it is still in the egg stage.

Dr Pickett uses special care to select chemicals that will do as little harm as possible to insect parasites and predators.

When we reach the point of using DDT, parathion, chlordane, and other new insecticides as routine control measures in the same way we have used the inorganic chemicals in the past, entomologists interested in biological control may as well throw in the sponge,

he says. Instead of these highly toxic, broad-spectrum insecticides, he places chief reliance on ryania (derived from ground stems of a tropical plant), nicotine sulphate, and lead arsenate. In certain situations very weak concentrations of DDT or malathion are used (1 or 2 ounces per 100 gallons – in contrast to the usual 1 or 2 pounds per 100 gallons). Although these two are the least toxic of the modern insecticides, Dr Pickett hopes by further research to replace them with safer and more selective materials.

How well has this programme worked? Nova Scotia orchardists who are following Dr Pickett's modified spray programme are producing as high a proportion of first-grade fruit as are those who are using intensive chemical applications. They are also getting as good production. They are getting these results, moreover, at a substantially lower cost. The outlay for insecticides in Nova Scotia apple orchards is only from 10 to 20 per cent of the amount spent in most other apple-growing areas.

More important than even these excellent results is the fact that the modified programme worked out by these Nova Scotian entomologists is not doing violence to Nature's balance. It is well on the way to realizing the philosophy stated by the Canadian entomologist G. C. Ullyett a decade ago:

We must change our philosophy, abandon our attitude of human superiority and admit that in many cases in natural environments we find ways and means of limiting populations of organisms in a more economical way than we can do it ourselves.

Chapter 16

The Rumblings of
an Avalanche

If Darwin were alive today the insect world would delight and astound him with its impressive verification of his theories of the survival of the fittest. Under the stress of intensive chemical spraying the weaker members of the insect populations are being weeded out. Now, in many areas and among many species only the strong and fit remain to defy our efforts to control them.

Nearly half a century ago, a professor of entomology at Washington State College, A. L. Melander, asked the now purely rhetorical question, 'Can insects become resistant to sprays?' If the answer seemed to Melander unclear, or slow in coming, that was only because he asked his question too soon – in 1914 instead of 40 years later. In the pre-DDT era, inorganic chemicals, applied on a scale that today would seem extraordinarily modest, produced here and there strains of insects that could survive chemical spraying or dusting. Melander himself had run into difficulty with the San José scale, for some years satisfactorily controlled by spraying with lime sulphur. Then in the Clarkson area of Washington the insects became refractory – they were harder to kill than in the orchards of the Wenatchee and Yakima valleys and elsewhere.

Suddenly the scale insects in other parts of the country seemed to have got the same idea: it was not necessary for them to die under the sprayings of lime sulphur, diligently and liberally applied by orchardists. Throughout much of the Mid-west thousands of acres of fine orchards were destroyed by insects now impervious to spraying.

Then in California the time-honoured method of placing canvas tents over trees and fumigating them with hydrocyanic acid began to yield disappointing results in certain areas, a problem that led to research at the California Citrus Experiment

Station, beginning about 1915 and continuing for a quarter of a century. Another insect to learn the profitable way of resistance was the codling moth, or appleworm, in the 1920s, although lead arsenate had been used successfully against it for some 40 years.

But it was the advent of DDT and all its many relatives that ushered in the true Age of Resistance. It need have surprised no one with even the simplest knowledge of insects or of the dynamics of animal populations that within a matter of a very few years an ugly and dangerous problem had clearly defined itself. Yet awareness of the fact that insects possess an effective counter-weapon to aggressive chemical attack seems to have dawned slowly. Only those concerned with disease-carrying insects seem by now to have been thoroughly aroused to the alarming nature of the situation; the agriculturists still for the most part blithely put their faith in the development of new and ever more toxic chemicals, although the present difficulties have been born of just such specious reasoning.

If understanding of the phenomenon of insect resistance developed slowly, it was far otherwise with resistance itself. Before 1945 only about a dozen species were known to have developed resistance to any of the pre-DDT insecticides. With the new organic chemicals and new methods for their intensive application, resistance began a meteoric rise that reached the alarming level of 137 species in 1960. No one believes the end is in sight. More than 1,000 technical papers have now been published on the subject. The World Health Organization has enlisted the aid of some 300 scientists in all parts of the world, declaring that 'resistance is at present the most important single problem facing vector-control programmes'. A distinguished British student of animal populations, Dr Charles Elton, has said, 'We are hearing the early rumblings of what may become an avalanche in strength.'

Sometimes resistance develops so rapidly that the ink is scarcely dry on a report hailing successful control of a species with some specified chemical when an amended report has to be issued. In South Africa, for example, cattlemen had long been plagued by the blue tick, from which, on one ranch alone, 600 head of cattle had died in one year. The tick had for some years been resistant to arsenical dips. Then benzene hexachloride was tried, and for a very short time all seemed to be well. Reports issued early in the year 1949 declared that the arsenic-resistant ticks could be controlled readily with the new chemical; later in the same year, a bleak notice of developing resistance had to be published. The situation prompted a writer in the *Leather Trades Review* to comment in 1950:

News such as this quietly trickling through scientific circles and appearing in small sections of the overseas press is enough to make headlines as big as those concerning the new atomic bomb if only the significance of the matter were properly understood.

Although insect resistance is a matter of concern in agriculture and forestry, it is in the field of public health that the most serious apprehensions have been felt. The relation between various insects and many diseases of man is an ancient one. Mosquitoes of the genus *Anopheles* may inject into the human bloodstream the single-celled organism of malaria. Other mosquitoes transmit yellow fever. Still others carry encephalitis. The housefly, which does not bite, nevertheless by contact may contaminate human food with the bacillus of dysentery, and in many parts of the world may play an important part in the transmission of eye diseases. The list of diseases and their insect carriers, or vectors, includes typhus and body lice, plague and rat fleas, African sleeping sickness and tsetse flies, various fevers and ticks, and innumerable others.

These are important problems and must be met. No responsible person contends that insect-borne disease should be ignored. The question that has now urgently presented itself is whether it is either wise or responsible to attack the problem by methods that are rapidly making it worse. The world has heard much of the triumphant war against disease through the control of insect vectors of infection, but it has heard little of the other side of the story – the defeats, the short-lived triumphs that now strongly support the alarming view that the insect enemy has been made actually stronger by our efforts. Even worse, we may have destroyed our very means of fighting.

A distinguished Canadian entomologist, Dr A. W. A. Brown,

was engaged by the World Health Organization to make a comprehensive survey of the resistance problem. In the resulting monograph, published in 1958, Dr Brown has this to say:

Barely a decade after the introduction of the potent synthetic insecticides in public health programmes, the main technical problem is the development of resistance to them by the insects they formerly controlled.

In publishing his monograph, the World Health Organization warned that

the vigorous offensive now being pursued against arthropod-borne diseases such as malaria, typhus fever, and plague risks a serious setback unless this new problem can be rapidly mastered.

What is the measure of this setback? The list of resistant species now includes practically all of the insect groups of medical importance. Apparently the blackflies, sand flies, and tsetse flies have not yet become resistant to chemicals. On the other hand, resistance among houseflies and body lice has now developed on a global scale. Malaria programmes are threatened by resistance among mosquitoes. The oriental rat flea, the principal vector of plague, has recently demonstrated resistance to DDT, a most serious development. Countries reporting resistance among a large number of other species represent every continent and most of the island groups.

Probably the first medical use of modern insecticides occurred in Italy in 1943 when the Allied Military Government launched a successful attack on typhus by dusting enormous numbers of people with DDT. This was followed two years later by extensive application of residual sprays for the control of malaria mosquitoes. Only a year later the first signs of trouble appeared. Both houseflies and mosquitoes of the genus *Culex* began to show resistance to the sprays. In 1948 a new chemical, chlordane, was tried as a supplement to DDT. This time good control was obtained for two years, but by August of 1950 chlordane-resistant flies appeared, and by the end of that year all of the houseflies as well as the *Culex* mosquitoes seemed to be resistant to chlordane. As rapidly as new chemicals were brought into use, resistance developed. By the end of 1951, DDT, methoxychlor, chlordane, heptachlor, and benzene hexachloride had joined the list of chemicals no longer effective. The flies, mean-while, had become 'fantastically abundant'.

The same cycle of events was being repeated in Sardinia during the late 1940s. In Denmark, products containing DDT were first used in 1944; by 1947 fly control had failed in many places. In some areas of Egypt, flies had already become resistant to DDT by 1948; BHC was substituted but was effective for less than a year. One Egyptian village in particular symbolizes the problem. Insecticides gave good control of flies in 1950 and during this same year the infant mortality rate was reduced by nearly 50 per cent. The next year, nevertheless, flies were resistant to DDT and chlordane. The fly population returned to its former level; so did infant mortality.

In the United States, DDT resistance among flies had become widespread in the Tennessee Valley by 1948. Other areas followed. Attempts to restore control with dieldrin met with little success, for in some places the flies developed strong resistance to this chemical *within only two months*. After running through all the available chlorinated hydrocarbons, control agencies turned to the organic phosphates, but here again the story of resistance was repeated. The present conclusion of experts is that 'housefly control has escaped insecticidal techniques and once more must be based on general sanitation'.

The control of body lice in Naples was one of the earliest and most publicized achievements of DDT. During the next few years its success in Italy was matched by the successful control of lice affecting some two million people in Japan and Korea in the winter of 1945–6. Some premonition of trouble ahead might have been gained by the failure to control a typhus epidemic in Spain in 1948. Despite this failure in actual practice, encouraging laboratory experiments led entomologists to believe lice were unlikely to develop resistance. Events in Korea in the winter of 1950–1 were therefore startling. When DDT powder was applied to a group of Korean soldiers the extraordinary result was an actual increase in the infestation of lice. When lice were collected and tested, it was found that 5 per cent DDT powder caused no increase in their natural mortality rate. Similar results among lice collected from vagrants in Tokyo, from an asylum in Itabashi, and from refugee camps in Syria, Jordan, and eastern

Egypt, confirmed the ineffectiveness of DDT for the control of lice and typhus. When by 1957 the list of countries in which lice had become resistant to DDT was extended to include Iran, Turkey, Ethiopia, West Africa, South Africa, Peru, Chile, France, Yugoslavia, Afghanistan, Uganda, Mexico, and Tanganyika, the initial triumph in Italy seemed dim indeed.

The first malaria mosquito to develop resistance to DDT was *Anopheles sacharovi* in Greece. Extensive spraying was begun in 1946 with early success; by 1949, however, observers noticed that adult mosquitoes were resting in large numbers under road bridges, although they were absent from houses and stables that had been treated. Soon this habit of outside resting was extended to caves, outbuildings, and culverts and to the foliage and trunks of orange trees. Apparently the adult mosquitoes had become sufficiently tolerant of DDT to escape from sprayed buildings and rest and recover in the open. A few months later they were able to remain in houses, where they were found resting on treated walls.

This was a portent of the extremely serious situation that has now developed. Resistance to insecticides by mosquitoes of the anopheline group has surged upwards at an astounding rate, being created by the thoroughness of the very house-spraying programmes designed to eliminate malaria. In 1956, only 5 species of these mosquitoes displayed resistance; by early 1960 the number had risen from 5 to 28! The number includes very dangerous malaria vectors in West Africa, the Middle East, Central America, Indonesia, and the eastern European region.

Among the other mosquitoes, including carriers of other diseases, the pattern is being repeated. A tropical mosquito that carries parasites responsible for such diseases as elephantiasis has become strongly resistant in many parts of the world. In some areas of the United States the mosquito vector of western equine encephalitis has developed resistance. An even more serious problem concerns the vector of yellow fever, for centuries one of the great plagues of the world. Insecticide-resistant strains of this mosquito have occurred in South-East Asia and are now common in the Caribbean region.

The consequences of resistance in terms of malaria and other diseases are indicated by reports from many parts of the world.

An outbreak of yellow fever in Trinidad in 1954 followed failure to control the vector mosquito because of resistance. There has been a flare-up of malaria in Indonesia and Iran. In Greece, Nigeria, and Liberia the mosquitoes continue to harbour and transmit the malaria parasite. A reduction of diarrhœal disease achieved in Georgia through fly control was wiped out within about a year. The reduction in acute conjunctivitis in Egypt, also attained through temporary fly control, did not last beyond 1950.

Less serious in terms of human health, but vexatious as man measures economic values, is the fact that salt-marsh mosquitoes in Florida also are showing resistance. Although these are not vectors of disease, their presence in bloodthirsty swarms had rendered large areas of coastal Florida uninhabitable until control – of an uneasy and temporary nature – was established. But this was quickly lost.

The ordinary house mosquito is here and there developing resistance, a fact that should give pause to many communities that now regularly arrange for wholesale spraying. This species is now resistant to several insecticides, among which is the almost universally used DDT, in Italy, Israel, Japan, France, and parts of the United States, including California, Ohio, New Jersey, and Massachusetts.

Ticks are another problem. The woodtick, vector of spotted fever, has recently developed resistance; in the brown dog tick the ability to escape a chemical death has long been thoroughly and widely established. This poses problems for human beings as well as for dogs. The brown dog tick is a semi-tropical species and when it occurs as far north as New Jersey it must live over winter in heated buildings rather than out of doors. John C. Pallister of the American Museum of Natural History reported in the summer of 1959 that his department had been getting a number of calls from neighbouring apartments on Central Park West.

Every now and then [Mr Pallister said] a whole apartment house gets infested with young ticks, and they're hard to get rid of. A dog will pick up ticks in Central Park, and then the ticks lay eggs and they hatch in the apartment. They seem immune to DDT or chlordane or most of our modern sprays. It used to be very unusual to have ticks in New

York City, but now they're all over here and on Long Island, in Westchester and on up into Connecticut. We've noticed this particularly in the past five or six years.

The German cockroach throughout much of North America has become resistant to chlordane, once the favourite weapon of exterminators who have now turned to the organic phosphates. However, the recent development of resistance to these insecticides confronts the exterminators with the problem of where to go next.

Agencies concerned with vector-borne disease are at present coping with their problems by switching from one insecticide to another as resistance develops. But this cannot go on indefinitely, despite the ingenuity of the chemists in supplying new materials. Dr Brown has pointed out that we are travelling 'a one-way street'. No one knows how long the street is. If the dead end is reached before control of disease-carrying insects is achieved, our situation will indeed be critical.

With insects that infest crops the story is the same.

To the list of about a dozen agricultural insects showing resistance to the inorganic chemicals of an earlier era there is now added a host of others resistant to DDT, BHC, lindane, toxaphene, dieldrin, aldrin, and even to the phosphates from which so much was hoped. The total number of resistant species among crop-destroying insects had reached 65 in 1960.

The first cases of DDT resistance among agricultural insects appeared in the United States in 1951, about six years after its first use. Perhaps the most troublesome situation concerns the codling moth, which is now resistant to DDT in practically all of the world's apple-growing regions. Resistance in cabbage insects is creating another serious problem. Potato insects are escaping chemical control in many sections of the United States. Six species of cotton insects, along with an assortment of thrips, fruit moths, leaf hoppers, caterpillars, mites, aphids, wireworms, and many others now are able to ignore the farmer's assault with chemical sprays.

The chemical industry is perhaps understandably loath to face up to the unpleasant fact of resistance. Even in 1959, with more than 100 major insect species showing definite resistance to chemicals, one of the leading journals in the field of agricultural chemistry spoke of 'real or imagined' insect resistance. Yet hopefully as the industry may turn its face the other way, the problem simply does not go away, and it presents some unpleasant economic facts. One is that the cost of insect control by chemicals is increasing steadily. It is no longer possible to stockpile materials well in advance; what today may be the most promising of insecticidal chemicals may be the dismal failure of tomorrow. The very substantial financial investment involved in backing and launching an insecticide may be swept away as the insects prove once more that the effective approach to nature is not through brute force. And however rapidly technology may invent new uses for insecticides and new ways of applying them, it is likely to find the insects keeping a lap ahead.

Darwin himself could scarcely have found a better example of the operation of natural selection than is provided by the way the mechanism of resistance operates. Out of an original population, the members of which vary greatly in qualities of structure, behaviour, or physiology, it is the 'tough' insects that survive chemical attack. Spraying kills off the weaklings. The only survivors are insects that have some inherent quality that allows them to escape harm. These are the parents of the new generation, which, by simple inheritance, possesses all the qualities of 'toughness' inherent in its forebears. Inevitably it follows that intensive spraying with powerful chemicals only makes worse the problem it is designed to solve. After a few generations, instead of a mixed population of strong and weak insects, there results a population consisting entirely of tough, resistant strains.

The means by which insects resist chemicals probably vary and as yet are not thoroughly understood. Some of the insects that defy chemical control are thought to be aided by a structural advantage, but there seems to be little actual proof of this. That immunity exists in some strains is clear, however, from observations like those of Dr Briejèr, who reports watching flies at the Pest Control Institute at Springforbi, Denmark, 'disporting themselves in DDT as much at home as primitive sorcerers cavorting over red-hot coals'.

Similar reports come from other parts of the world. In Malaya, at Kuala Lumpur, mosquitoes at first reacted to DDT by leaving the treated interiors. As resistance developed, however, they could be found at rest on surfaces where the deposit of DDT beneath them was clearly visible by torchlight. And in an army camp in southern Taiwan samples of resistant bedbugs were found actually carrying a deposit of DDT powder on their bodies. When these bedbugs were experimentally placed in cloth impregnated with DDT, they lived for as long as a month; they proceeded to lay their eggs; and the resulting young grew and thrived.

Nevertheless, the quality of resistance does not necessarily depend on physical structure. DDT-resistant flies possess an enzyme that allows them to detoxify the insecticide to the less toxic chemical DDE. This enzyme occurs only in flies that possess a genetic factor for DDT resistance. This factor is, of course, hereditary. How flies and other insects detoxify the organic phosphorus chemicals is less clearly understood.

Some behavioural habit may also place the insect out of reach of chemicals. Many workers have noticed the tendency of resistant flies to rest more on untreated horizontal surfaces than on treated walls. Resistant houseflies may have the stable-fly habit of sitting still in one place, thus greatly reducing the frequency of their contact with residues of poison. Some malaria mosquitoes have a habit that so reduces their exposure to DDT as to make them virtually immune. Irritated by the spray, they leave the huts and survive outside.

Ordinarily resistance takes two or three years to develop, although occasionally it will do so in only one season, or even less. At the other extreme it may take as long as six years. The number of generations produced by an insect population in a year is important, and this varies with species and climate. Flies in Canada, for example, have been slower to develop resistance than those in southern United States, where long hot summers favour a rapid rate of reproduction.

The hopeful question is sometimes asked, 'If insects can become resistant to chemicals, could human beings do the same thing?' Theoretically they could; but since this would take hundreds or even thousands of years, the comfort to those living now is slight. Resistance is not something that develops in an individual. If he possesses at birth some qualities that make him less susceptible than others to poisons he is more likely to survive and produce children. Resistance, therefore, is something that develops in a population after time measured in several or many generations. Human populations reproduce at the rate of roughly three generations per century, but new insect generations arise in a matter of days or weeks.

It is more sensible in some cases to take a small amount of damage in preference to having none for a time but paying for it in the long run by losing the very means of fighting [is the advice given in Holland by Dr Briejèr in his capacity as director of the Plant Protection Service]. Practical advice should be 'Spray as little as you possibly can' rather than 'Spray to the limit of your capacity'.... Pressure on the pest population should always be as slight as possible.

Unfortunately, such vision has not prevailed in the corresponding agricultural services of the United States. The Department of Agriculture's *Yearbook* for 1952, devoted entirely to insects, recognizes the fact that insects become resistant but says, 'More applications or greater quantities of the insecticides are needed then for adequate control.' The Department does not say what will happen when the only chemicals left untried are those that render the earth not only insectless but lifeless. But in 1959, only seven years after this advice was given, a Connecticut entomologist was quoted in the *Journal of Agricultural and Food Chemistry* to the effect that on at least one or two insect pests *the last available* new material was then being used.

Dr Briejèr says:

It is more than clear that we are travelling a dangerous road.... *We are going to have to do some very energetic research on other control measures, measures that will have to be biological, not chemical. Our aim should be to guide natural processes as cautiously as possible in the desired direction rather than to use brute force....*

We need a more high-minded orientation and a deeper insight, which I miss in many researchers. Life is a miracle beyond our comprehension, and we should reverence it even where we have to struggle against it.... The resort to weapons such as insecticides to control it is a proof of insufficient knowledge and of an incapacity so to guide the processes of nature that brute force becomes unnecessary. Humbleness is in order; there is no excuse for scientific conceit here.

Chapter 17

The Other Road

We stand now where two roads diverge. But unlike the roads in Robert Frost's familiar poem, they are not equally fair. The road we have long been travelling is deceptively easy, a smooth super-highway on which we progress with great speed, but at its end lies disaster. The other fork of the road – the one 'less travelled by' – offers our last, our only chance to reach a destination that assures the preservation of our earth.

The choice, after all, is ours to make. If, having endured much, we have at last asserted our 'right to know', and if, knowing, we have concluded that we are being asked to take senseless and frightening risks, then we should no longer accept the counsel of those who tell us that we must fill our world with poisonous chemicals; we should look about and see what other course is open to us.

A truly extraordinary variety of alternatives to the chemical control of insects is available. Some are already in use and have achieved brilliant success. Others are in the stage of laboratory testing. Still others are little more than ideas in the minds of imaginative scientists, waiting for the opportunity to put them to the test. All have this in common: they are *biological* solutions, based on understanding of the living organisms they seek to control, and of the whole fabric of life to which these organisms belong. Specialists representing various areas of the vast field of biology are contributing – entomologists, pathologists, geneticists, physiologists, biochemists, ecologists – all pouring their knowledge and their creative inspirations into the formation of a new science of biotic controls.

Any science may be likened to a river [says a Johns Hopkins biologist, Professor Carl P. Swanson]. It has its obscure and unpretentious beginning; its quiet stretches as well as its rapids; its periods of drought as well as of fullness. It gathers momentum with the work of many investigators and as it is fed by other streams of thought; it is deepened and broadened by the concepts and generalizations that are gradually evolved.

So it is with the science of biological control in its modern sense. In America it had its obscure beginnings a century ago with the first attempts to introduce natural enemies of insects that were proving troublesome to farmers, an effort that sometimes moved slowly or not at all, but now and again gathered speed and momentum under the impetus of an outstanding success. It had its period of drought when workers in applied entomology, dazzled by the spectacular new insecticides of the 1940s, turned their backs on all biological methods and set foot on 'the treadmill of chemical control'. But the goal of an insect-free world continued to recede. Now at last, as it has become apparent that the heedless and unrestrained use of chemicals is a greater menace to ourselves than to the targets, the river which is the science of biotic control flows again, fed by new streams of thought.

Some of the most fascinating of the new methods are those that seek to turn the strength of a species against itself – to use the drive of an insect's life forces to destroy it. The most spectacular of these approaches is the 'male sterilization' technique developed by the chief of the United States Department of Agriculture's Entomology Research Branch, Dr Edward Knipling, and his associates.

About a quarter of a century ago Dr Knipling startled his colleagues by proposing a unique method of insect control. If it were possible to sterilize and release large numbers of insects, he theorized, the sterilized males would, under certain conditions, compete with the normal wild males so successfully that, after repeated releases, only infertile eggs would be produced and the population would die out.

The proposal was met with bureaucratic inertia and with scepticism from scientists, but the idea persisted in Dr Knipling's mind. One major problem remained to be solved before it could be put to the test – a practical method of insect sterilization had to be found. Academically, the fact that insects could be sterilized by exposure to X-ray had been known since 1916, when an entomologist by the name of G. A. Runner reported such sterilization of cigarette beetles. Hermann Muller's pioneering work on the production of mutations by X-ray opened up vast new areas of thought in the late 1920s, and by the middle of the century various workers had reported the sterilization by X-rays or gamma rays of at least a dozen species of insects.

But these were laboratory experiments, still a long way from practical application. About 1950, Dr Knipling launched a serious effort to turn insect sterilization into a weapon that would wipe out a major insect enemy of livestock in the South, the screw-worm fly. The females of this species lay their eggs in any open wound of a warm-blooded animal. The hatching larvae are parasitic, feeding on the flesh of the host. A full-grown steer may succumb to a heavy infestation in 10 days, and livestock losses in the United States have been estimated at $40,000,000 a year. The toll of wildlife is harder to measure, but it must be great. Scarcity of deer in some areas of Texas is attributed to the screw-worm. This is a tropical or subtropical insect, inhabiting South and Central America and Mexico, and in the United States normally restricted to the South-west. About 1933, however, it was accidentally introduced into Florida, where the climate allowed it to survive over winter and to establish populations. It even pushed into southern Alabama and Georgia, and soon the livestock industry of the south-eastern states was faced with annual losses running to $20,000,000.

A vast amount of information on the biology of the screw-worm had been accumulated over the years by Agriculture Department scientists in Texas. By 1954, after some preliminary field trials on Florida islands, Dr Knipling was ready for a full-scale test of his theory. For this, by arrangement with the Dutch Government, he went to the island of Curaçao in the Caribbean, cut off from the mainland by at least 50 miles of sea.

Beginning in August 1954, screw-worms reared and sterilized in an Agriculture Department laboratory in Florida were flown to Curaçao and released from aeroplanes at the rate of about 400 per square mile per week. Almost at once the number of egg masses deposited on experimental goats began to decrease, as did their fertility. Only seven weeks after the releases were started, all eggs were infertile. Soon it was impossible to find a single egg mass, sterile or otherwise. The screw-worm had indeed been eradicated on Curaçao.

The resounding success of the Curaçao experiment whetted the appetites of Florida livestock raisers for a similar feat that would relieve them of the scourge of screw-worms. Although the difficulties here were relatively enormous – an area 300 times as large as the small Caribbean island – in 1957 the United States Department of Agriculture and the State of Florida joined in providing funds for an eradication effort. The project involved the weekly production of about 50 million screw-worms at a specially constructed 'fly factory', the use of 20 light aeroplanes to fly prearranged flight patterns, five to six hours daily, each plane carrying a thousand paper cartons, each carton containing 200 to 400 irradiated flies.

The cold winter of 1957–8, when freezing temperatures gripped northern Florida, gave an unexpected opportunity to start the programme while the screw-worm populations were reduced and confined to a small area. By the time the programme was considered complete at the end of 17 months, $3\frac{1}{2}$ billion artificially reared, sterilized flies had been released over Florida and sections of Georgia and Alabama. The last-known animal wound infestation that could be attributed to screw-worms occurred in February 1959. In the next few weeks several adults were taken in traps. Thereafter no trace of the screw-worm could be discovered. Its extinction in the South-east had been accomplished – a triumphant demonstration of the worth of scientific creativity, aided by thorough basic research, persistence, and determination.

Now a quarantine barrier in Mississippi seeks to prevent the re-entrance of the screw-worm from the South-west, where it is firmly entrenched. Eradication there would be a formidable undertaking, considering the vast areas involved and the probability of re-invasion from Mexico. Nevertheless, the stakes are high and the thinking in the Department seems to be that some sort of programme, designed at least to hold the screw-worm populations at very low levels, may soon be attempted in Texas and other infested areas of the South-west.

The brilliant success of the screw-worm campaign has stimulated tremendous interest in applying the same methods to other insects. Not all, of course, are suitable subjects for this technique, much depending on details of the life history, population density, and reactions to radiation.

Experiments have been undertaken by the British in the hope that the method could be used against the tsetse fly in Rhodesia. This insect infests about a third of Africa, posing a menace to human health and preventing the keeping of livestock in an area of some 40 million square miles of wooded grasslands. The habits of the tsetse differ considerably from those of the screw-worm fly, and although it can be sterilized by radiation some technical difficulties remain to be worked out before the method can be applied.

The British have already tested a large number of other species for susceptibility to radiation. United States scientists have had some encouraging early results with the melon fly and the oriental and Mediterranean fruit flies in laboratory tests in Hawaii and field tests on the remote island of Rota. The corn borer and the sugarcane borer are also being tested. There are possibilities, too, that insects of medical importance might be controlled by sterilization. A Chilean scientist has pointed out that malaria-carrying mosquitoes persist in his country in spite of insecticide treatment; the release of sterile males might then provide the final blow needed to eliminate this population.

The obvious difficulties of sterilizing by radiation have led to search for an easier method of accomplishing similar results, and there is now a strongly running tide of interest in chemical sterilants.

Scientists at the Department of Agriculture laboratory in Orlando, Florida, are now sterilizing the housefly in laboratory experiments and even in some field trials, using chemicals incorporated in suitable foods. In a test on an island in the Florida Keys in 1961, a population of flies was nearly wiped out within a period of only five weeks. Repopulation of course followed from nearby islands, but as a pilot project the test was successful. The Department's excitement about the promise of this method is easily understood. In the first place, as we have seen, the housefly has now become virtually uncontrollable by insecticides. A completely new method of control is undoubtedly needed. One of the problems of sterilization by radiation is that this requires not only artificial rearing but the release of sterile males in larger number than are present in the wild population. This could be done with the screw-worm, which is actually not an abundant insect. With the housefly, however, more than doubling the population through releases could be highly objectionable, even though the increase would be only temporary. A chemical sterilant, on the other hand, could be combined with a bait substance and introduced into the natural environment of the fly; insects feeding on it would become sterile and in the course of time the sterile flies would predominate and the insects would breed themselves out of existence.

The testing of chemicals for a sterilizing effect is much more difficult than the testing of chemical poisons. It takes 30 days to evaluate one chemical – although, of course, a number of tests can be run concurrently. Yet between April 1958 and December 1961 several hundred chemicals were screened at the Orlando laboratory for a possible sterilizing effect. The Department of Agriculture seems happy to have found among these even a handful of chemicals that show promise.

Now other laboratories of the Department are taking up the problem, testing chemicals against stable flies, mosquitoes, boll weevils, and an assortment of fruit flies. All this is at present experimental but in the few years since work began on chemosterilants the project has grown enormously. In theory it has many attractive features. Dr Knipling has pointed out that effective chemical insect sterilization 'might easily outdo some of the best of known insecticides'. Take an imaginary situation in which a population of a million insects is multiplying five times in each generation. An insecticide might kill 90 per cent of each generation, leaving 125,000 insects alive after the third generation. In contrast, a chemical that would produce 90 per cent sterility would leave only 125 insects alive.

On the other side of the coin is the fact that some extremely potent chemicals are involved. It is fortunate that at least during these early stages most of the men working with chemosterilants seem mindful of the need to find safe chemicals and safe methods of application. None the less, suggestions are heard here and there that these sterilizing chemicals might be applied as aerial sprays – for example, to coat the foliage chewed by gypsy moth larvae. To attempt any such procedure without thorough advance research on the hazards involved would be the height of irresponsibility. If the potential hazards of the chemosterilants are not constantly borne in mind we could easily find ourselves in even worse trouble than that now created by the insecticides.

The sterilants currently being tested fall generally into two groups, both of which are extremely interesting in their mode of action. The first are intimately related to the life processes, or metabolism, of the cell; i.e., they so closely resemble a substance the cell or tissue needs that the organism 'mistakes' them for the true metabolite and tries to incorporate them in its normal building processes. But the fit is wrong in some detail and the process comes to a halt. Such chemicals are called antimetabolites.

The second group consists of chemicals that act on the chromosomes, probably affecting the gene chemicals and causing the chromosomes to break up. The chemosterilants of this group are alkylating agents, which are extremely reactive chemicals, capable of intense cell destruction, damage to chromosomes, and production of mutations. It is the view of Dr Peter Alexander of the Chester Beatty Research Institute in London that 'any alkylating agent which is effective in sterilizing insects would also be a powerful mutagen and carcinogen'. Dr Alexander feels that any conceivable use of such chemicals in insect control would be 'open to the most severe objections'. It is to be hoped, therefore, that the present experiments will lead not to actual use of these particular chemicals but to the discovery of others that will be safe and also highly specific in their action on the target insect.

Some of the most interesting of the recent work is concerned with still other ways of forging weapons from the insect's own life processes. Insects produce a variety of venoms, attractants, repellants. What is the chemical nature of these secretions? Could we make use of them as, perhaps, very selective insecticides? Scientists at Cornell University and elsewhere are trying to find answers to some of these questions, studying the defence mechanisms by which many insects protect themselves from attack by predators, working out the chemical structure of insect secretions. Other scientists are working on the so-called 'juvenile hormone', a powerful substance which prevents metamorphosis of the larval insect until the proper stage of growth has been reached.

Perhaps the most immediately useful result of this exploration of insect secretion is the development of lures, or attractants. Here again, nature has pointed the way. The gypsy moth is an especially intriguing example. The female moth is too heavy-bodied to fly. She lives on or near the ground, fluttering about among low vegetation or creeping up tree trunks. The male, on the contrary, is a strong flier and is attracted even from consider-able distances by a scent released by the female from special glands. Entomologists have taken advantage of this fact for a good many years, laboriously preparing this sex attractant from the bodies of the female moths. It was then used in traps set for the males in census operations along the fringe of the insect's range. But this was an extremely expensive procedure. Despite the much publicized infestations in the north-eastern states, there were not enough gypsy moths to provide the material, and hand-collected female pupae had to be imported from Europe, sometimes at a cost of half a dollar per tip. It was a tremendous breakthrough, therefore, when, after years of effort, chemists of the Agriculture Department recently succeeded in isolating the attractant. Following upon this discovery was the successful pre-paration of a closely related synthetic material from a constituent of castor oil; this not only deceives the male moths but is appar-ently fully as attractive as the natural substance. As little as one microgram ($1/1000$ gram) in a trap is an effective lure.

All this is of much more than academic interest, for the new and economic 'gyplure' might be used not merely in census operations but in control work. Several of the more attractive possibilities are now being tested. In what might be termed an experiment in psychological warfare, the attractant is combined with a granular material and distributed by planes. The aim is to confuse the male moth and alter the normal behaviour so that, in the welter of attractive scents, he cannot find the true scent trail leading to the female. This line of attack is being carried even further in experiments aimed at deceiving the male into attempting to mate with a spurious female. In the laboratory, male gypsy moths have attempted copulation with chips of wood, vermiculite, and other small, inanimate objects, so long as they were suitably impregnated with gyplure. Whether such diversion of the mating instinct into non-productive channels would actually serve to reduce the population remains to be tested, but it is an interesting possibility.

The gypsy moth lure was the first insect sex attractant to be synthesized, but probably there will soon be others. A number of agricultural insects are being studied for possible attractants that man could imitate. Encouraging results have been obtained with the Hessian fly and the tobacco hornworm.

Combinations of attractants and poisons are being tried against several insect species. Government scientists have developed an attractant called methyl-eugenol, which males of the oriental fruit fly and the melon fly find irresistible. This has been combined with a poison in tests in the Bonin Islands 450 miles south of Japan. Small pieces of fibreboard were impregnated with the two chemicals and were distributed by air over the entire island chain to attract and kill the male flies. This programme of 'male annihilation' was begun in 1960: a year later the Agriculture Department estimated that more than 99 per cent of the population had been eliminated. The method as here applied seems to have marked advantages over the conventional broadcasting of insecticides. The poison, an organic phosphorus chemical, is confined to squares of fibreboard, which are unlikely to be eaten by wildlife; its residues, moreover, are quickly dissipated and so are not potential contaminants of soil or water.

But not all communication in the insect world is by scents that lure or repel. Sound also may be a warning or an attraction. The constant stream of ultrasonic sound that issues from a bat in flight (serving as a radar system to guide it through darkness) is heard by certain moths, enabling them to avoid capture. The wing sounds of approaching parasitic flies warn the larvae of some sawflies to herd together for protection. On the other hand, the sounds made by certain wood-boring insects enable their parasites to find them, and to the male mosquito the wingbeat of the female is a siren song.

What use, if any, can be made of this ability of the insect to detect and react to sound? As yet in the experimental stage, but none the less interesting, is the initial success in attracting male mosquitoes to playback recordings of the flight sound of the female. The males were lured to a charged grid and so killed. The repellant effect of bursts of ultrasonic sound is being tested in Canada against corn borer and cutworm moths. Two authorities on animal sound, Professors Hubert and Mable Frings of the University of Hawaii, believe that a field method of influencing the behaviour of insects with sound only awaits discovery of the proper key to unlock and apply the vast existing knowledge of insect sound production and reception. Repellant sounds may offer greater possibilities than attractants. The Fringses are known for their discovery that starlings scatter in alarm before a recording of the distress cry of one of their fellows; perhaps somewhere in this fact is a central truth that may be applied to insects. To practical men of industry the possibilities seem real enough so that at least one major electronic corporation is preparing to set up a laboratory to test them.

Sound is also being tested as an agent of direct destruction. Ultrasonic sound will kill all mosquito larvae in a laboratory tank; however, it kills other aquatic organisms as well. In other experiments, blowflies, mealworms, and yellow-fever mosquitoes have been killed by airborne ultrasonic sound in a matter of seconds. All such experiments are first steps towards wholly new concepts of insect control which the miracles of electronics may some day make a reality.

The new biotic control of insects is not wholly a matter of electronics and gamma radiation and other products of man's inventive mind. Some of its methods have ancient roots, based on the knowledge that, like ourselves, insects are subject to disease. Bacterial infections sweep through their populations like the plagues of old; under the onset of a virus their hordes sicken and die. The occurrence of disease in insects was known before the time of Aristotle; the maladies of the silkworm were celebrated in medieval poetry; and through study of the diseases of this same insect the first understanding of the principles of infectious disease came to Pasteur.

Insects are beset not only by viruses and bacteria but also by fungi, protozoa, microscopic worms, and other beings from all that unseen world of minute life that, by and large, befriends mankind. For the microbes include not only disease organisms but those that destroy waste matter, make soils fertile, and enter into countless biological processes like fermentation and nitrification. Why should they not also aid us in the control of insects?

One of the first to envision such use of micro-organisms was the nineteenth-century zoologist Elie Metchnikoff. During the concluding decades of the nineteenth and the first half of the twentieth centuries the idea of microbial control was slowly taking form. The first conclusive proof that an insect could be brought under control by introducing a disease into its environment came in the late 1930s with the discovery and use of milky disease for the Japanese beetle, which is caused by the spores of a bacterium belonging to the genus *Bacillus*. This classic example of bacterial control has a long history of use in the eastern part of the United States, as I have pointed out in Chapter 7.

High hopes now attend tests of another bacterium of this genus – *Bacillus thuringiensis* – originally discovered in Germany in 1911 in the province of Thuringia, where it was found to cause a fatal septicaemia in the larvae of the flour moth. This bacterium actually kills by poisoning rather than by disease. Within its vegetative rods there are formed, along with spores, peculiar crystals composed of a protein substance highly toxic to certain insects, especially to the larvae of the moth-like lepidopteras. Shortly after eating foliage coated with this toxin the larva suffers paralysis, stops feeding, and soon dies. For practical purposes, the fact that feeding is interrupted promptly is of course an enormous advantage, for crop damage stops almost as soon as the pathogen is applied. Compounds containing spores of *Bacillus thuringiensis* are now being manufactured by several firms in the United States under various trade names. Field tests are being made in several countries: in France and Germany against larvae of the cabbage butterfly, in Yugoslavia against the autumn webworm, in the Soviet Union against a tent caterpillar. In Panama, where tests were begun in 1961, this bacterial insecticide may be the answer to one or more of the serious problems confronting banana growers. There the root borer is a serious pest of the banana, so weakening its roots that the trees are easily toppled by wind. Dieldrin has been the only chemical effective against the borer, but it has now set in motion a chain of disaster.

The borers are becoming resistant. The chemical has also destroyed some important insect predators and so has caused an increase in the tortricids – small, stout-bodied moths whose larvae scar the surface of the bananas. There is reason to hope the new microbial insecticide will eliminate both the tortricids and the borers and that it will do so without upsetting natural controls.

In eastern forests of Canada and the United States bacterial insecticides may be one important answer to the problems of such forest insects as the budworms and the gypsy moth. In 1960 both countries began field tests with a commercial preparation of *Bacillus thuringiensis*. Some of the early results have been encouraging. In Vermont, for example, the end results of bacterial control were as good as those obtained with DDT. The main technical problem now is to find a carrying solution that will stick the bacterial spores to the needles of the evergreens. On crops this is not a problem – even a dust can be used. Bacterial insecticides have already been tried on a wide variety of vegetables, especially in California.

Meanwhile, other perhaps less spectacular work is concerned with viruses. Here and there in California fields of young alfalfa are being sprayed with a substance as deadly as any insecticide for the destructive alfalfa caterpillar – a solution containing a virus obtained from the bodies of caterpillars that have died because of infection with this exceedingly virulent disease. The bodies of only five diseased caterpillars provide enough virus to treat an acre of alfalfa. In some Canadian forests a virus that affects pine sawflies has proved so effective in control that it has replaced insecticides.

Scientists in Czechoslovakia are experimenting with protozoa against webworms and other insect pests, and in the United States a protozoan parasite has been found to reduce the egg-laying potential of the corn borer.

To some the term microbial insecticide may conjure up pictures of bacterial warfare that would endanger other forms of life. This is not true. In contrast to chemicals, insect pathogens are harmless to all but their intended targets. Dr Edward Steinhaus, an outstanding authority on insect pathology, has stated emphatically that there is no authenticated recorded instance of a true insect pathogen having caused an infectious disease in a vertebrate animal either experimentally or in nature.

The insect pathogens are so specific that they infect only a small group of insects – sometimes a single species. Biologically they do not belong to the type of organisms that cause disease in higher animals or in plants. Also, as Dr Steinhaus points out, outbreaks of insect disease in nature always remain confined to insects, affecting neither the host plants nor animals feeding on them.

Insects have many natural enemies – not only microbes of many kinds but other insects. The first suggestion that an insect might be controlled by encouraging its enemies is generally credited to Erasmus Darwin about 1800. Probably because it was the first generally practised method of biological control, this setting of one insect against another is widely but erroneously thought to be the only alternative to chemicals.

In the United States the true beginnings of conventional biological control date from 1888 when Albert Koebele, the first of a growing army of entomologist explorers, went to Australia to search for natural enemies of the cottony cushion scale that threatened the California citrus industry with destruction. As we have seen in Chapter 15, the mission was crowned with spectacular success, and in the century that followed the world has been combed for natural enemies to control the insects that have come uninvited to our shores. In all, about 100 species of imported predators and parasites have become established. Besides the vedalia beetles brought in by Koebele, other importations have been highly successful. A wasp imported from Japan established complete control of an insect attacking Eastern apple orchards. Several natural enemies of the spotted alfalfa aphid, an accidental import from the Middle East, are credited with saving the California alfalfa industry. Parasites and predators of the gypsy moth achieved good control, as did the *Tiphia* wasp against the Japanese beetle. Biological control of scales and mealy bugs is estimated to save California several millions of dollars a year – indeed, one of the leading entomologists of that state, Dr Paul DeBach, has estimated that for an investment of $4,000,000 in biological control work California has received a return of $100,000,000.

Examples of successful biological control of serious pests by importing their natural enemies are to be found in some 40 countries distributed over much of the world. The advantages of such control over chemicals are obvious: it is relatively inexpensive, it is permanent, it leaves no poisonous residues. Yet biological control has suffered from lack of support. California is virtually alone among the states in having a formal programme in biological control, and many states have not even one entomologist who devotes full time to it. Perhaps for want of support biological control through insect enemies has not always been carried out with the scientific thoroughness it requires – exacting studies of its impact on the populations of insect prey have seldom been made, and releases have not always been made with the precision that might spell the difference between success and failure.

The predator and the preyed upon exist not alone, but as part of a vast web of life, all of which needs to be taken into account. Perhaps the opportunities for the more conventional types of biological control are greatest in the forests. The farmlands of modern agriculture are highly artificial, unlike anything nature ever conceived. But the forests are a different world, much closer to natural environments. Here, with a minimum of help and a maximum of non-interference from man, nature can have her way, setting up all that wonderful and intricate system of checks and balances that protects the forest from undue damage by insects.

In the United States our foresters seem to have thought of biological control chiefly in terms of introducing insect parasites and predators. The Canadians take a broader view, and some of the Europeans have gone farthest of all to develop the science of 'forest hygiene' to an amazing extent. Birds, ants, forest spiders, and soil bacteria are as much a part of a forest as the trees, in the view of European foresters, who take care to inoculate a new forest with these protective factors. The encouragement of birds is one of the first steps. In the modern era of intensive forestry the old hollow trees are gone and with them homes for woodpeckers and other tree-nesting birds. This lack is met by nesting boxes, which draw the birds back into the forest. Other boxes are specially designed for owls and for bats, so that these creatures may take over in the dark hours the work of insect hunting performed in daylight by the small birds.

But this is only the beginning. Some of the most fascinating control work in European forests employs the forest red ant as an aggressive insect predator – a species which, unfortunately, does not occur in North America. About 25 years ago Professor Karl Gösswald of the University of Würzburg developed a method of cultivating this ant and establishing colonies. Under his direction more than 10,000 colonies of the red ant have been established in about 90 test areas in the German Federal Republic. Dr Gösswald's method has been adopted in Italy and other countries, where ant farms have been established to supply colonies for distribution in the forests. In the Apennines, for example, several hundred nests have been set out to protect reforested areas.

'Where you can obtain in your forest a combination of birds' and ants' protection together with some bats and owls, the biological equilibrium has already been essentially improved,' says Dr Heinz Ruppertshofen, a forestry officer in Mölln, Germany, who believes that a single introduced predator or parasite is less effective than an array of the 'natural companions' of the trees.

New ant colonies in the forests at Mölln are protected from woodpeckers by wire netting to reduce the toll. In this way the woodpeckers, which have increased by 400 per cent in 10 years in some of the test areas, do not seriously reduce the ant colonies, and pay handsomely for what they take by picking harmful caterpillars off the trees. Much of the work of caring for the ant colonies (and the birds' nesting boxes as well) is assumed by a youth corps from the local school, children 10 to 14 years old. The costs are exceedingly low; the benefits amount to permanent protection of the forests.

Another extremely interesting feature of Dr Ruppertshofen's work is his use of spiders, in which he appears to be a pioneer. Although there is a large literature on the classification and natural history of spiders, it is scattered and fragmentary and deals not at all with their value as an agent of biological control. Of the 22,000 known kinds of spiders, 760 are native to Germany (and about 2,000 to the United States). Twenty-nine families of spiders inhabit German forests.

To a forester the most important fact about a spider is the kind of net it builds. The wheel-net spiders are most important, for the webs of some of them are so narrow-meshed that they can catch all flying insects. A large web (up to 16 inches in diameter) of the cross spider bears some 120,000 adhesive nodules on its strands. A single spider may destroy in her life of 18 months an average of 2,000 insects. A biologically sound forest has 50 to 150 spiders to the square metre (a little more than a square yard). Where there are fewer, the deficiency may be remedied by collecting and distributing the bag-like cocoons containing the eggs. 'Three cocoons of the wasp spider [which occurs also in America] yield a thousand spiders, which can catch 200,000 flying insects,' said Dr Ruppertshofen. The tiny and delicate young of the wheel-net spiders that emerge in the spring are especially important, he says, 'as they spin in a teamwork a net umbrella above the top shoots of the trees and thus protect the young shoots against the flying insects'. As the spiders moult and grow, the net is enlarged.

Canadian biologists have pursued rather similar lines of investigation, although with differences dictated by the fact that North American forests are largely natural rather than planted, and that the species available as aids in maintaining a healthy forest are somewhat different. The emphasis in Canada is on small mammals, which are amazingly effective in the control of certain insects, especially those that live within the spongy soil of the forest floor. Among such insects are the sawflies, so called because the female has a saw-shaped ovipositor with which she slits open the needles of evergreen trees in order to deposit her eggs. The larvae eventually drop to the ground and form cocoons in the peat of tamarack bogs or the duff under spruce or pines. But beneath the forest floor is a world honeycombed with the tunnels and runways of small mammals – whitefooted mice, voles, and shrews of various species. Of all these small burrowers, the voracious shrews find and consume the largest number of sawfly cocoons. They feed by placing a forefoot on the cocoon and biting off the end, showing an extraordinary ability to discriminate between sound and empty cocoons. And for their insatiable appetite the shrews have no rivals. Whereas a vole can consume about 200 cocoons a day, a shrew, depending on the species, may devour up to 800! This may result, according to laboratory tests, in destruction of 75 to 98 per cent of the cocoons present.

It is not surprising that the island of Newfoundland, which has no native shrews but is beset with sawflies, so eagerly desired some of these small, efficient mammals that in 1958 the introduction of the masked shrew – the most efficient sawfly predator – was attempted. Canadian officials report in 1962 that the attempt has been successful. The shrews are multiplying and are spreading out over the island, some marked individuals having been recovered as much as ten miles from the point of release.

There is, then, a whole battery of armaments available to the forester who is willing to look for permanent solutions that preserve and strengthen the natural relations in the forest. Chemical pest control in the forest is at best a stopgap measure bringing no real solution, at worst killing the fishes in the forest streams, bringing on plagues of insects, and destroying the natural controls and those we may be trying to introduce. By such violent measures, says Dr Ruppertshofen,

the partnership for life of the forest is entirely being unbalanced, and the catastrophes caused by parasites repeat in shorter and shorter periods... We, therefore, have to put an end to these unnatural manipulations brought into the most important and almost last natural living space which has been left for us.

Through all these new, imaginative, and creative approaches to the problem of sharing our earth with other creatures there runs a constant theme, the awareness that we are dealing with life – with living populations and all their pressures and counter-pressures, their surges and recessions. Only by taking account of such life forces and by cautiously seeking to guide them into channels favourable to ourselves can we hope to achieve a reasonable accommodation between the insect hordes and ourselves.

The current vogue for poisons has failed utterly to take into account these most fundamental considerations. As crude a weapon as the cave man's club, the chemical barrage has been hurled against the fabric of life – a fabric on the one hand delicate and destructible, on the other miraculously tough and resilient, and capable of striking back in unexpected ways. These extraordinary capacities of life have been ignored by the practitioners of chemical control who have brought to their task no 'high-minded orientation', no humility before the vast forces with which they tamper.

The 'control of nature' is a phrase conceived in arrogance, born of the Neanderthal age of biology and philosophy, when it was supposed that nature exists for the convenience of man. The concepts and practices of applied entomology for the most part date from that Stone Age of science. It is our alarming misfortune that so primitive a science has armed itself with the most modern and terrible weapons, and that in turning them against the insects it has also turned them against the earth.

Afterword

Very few books change the course of history. Those that have include Karl Marx's, *Das Capital*, Adam Smith's, *The Wealth of Nations*, Charles Darwin's, *The Origin of the Species* and, in the United States, Harriet Beecher Stowe's, *Uncle Tom's Cabin*. Rachel Carson's *Silent Spring* is another. It was serialized first in three successive issues of the *New Yorker* during the summer of 1962, published as a book in September, and chosen as a book-of-the-month club selection for October, where US Supreme Court Justice William O. Douglas called it 'the most revolutionary book since *Uncle Tom's Cabin*'. It initiated the contemporary environmental movement which undeniably influences the social policy of every nation. It was almost literally true, as one American editorial writer put it, that 'a few thousand words from Rachel Carson and the world took a new direction'.

Silent Spring was the final work of a creative writer and brilliant scientific synthesizer. Carson became bolder and angrier as she gathered her research and appraised the careless attitude of those who would spread millions of pounds of persistent chemical pesticides over the landscape without knowing the long-term impact of doing so.

Silent Spring translated the central truth of ecology: that everything in nature is related to everything else. It carefully explained how laying down a barrage of synthetic pesticides, and by analogy, the product of any other biological technology, might produce something different than the expected single outcome.

Carson's thinking was deeply influenced by the nuclear tensions of the Cold War period in which she worked. Politically sophisticated, a professional civil servant and

conservation advocate, she began her research at a time
when the rigid conformity demanded by the US government's anti-communist policies made even legitimate criticism of government policy risky. Science and technology
and those who worked in these fields were revered as the
saviours of the free world and the trustees of prosperity. In
Silent Spring Rachel Carson exposed these experts to public
scrutiny and made it clear that at best they had not done
their homework, and at worst they had withheld the truth.
She was one of the first to convince a complacent and
increasingly affluent post-war generation that the government could not be trusted to take care of them, and urged
individual citizens to assume responsibility for understanding the impact of government policies, and to challenge
those that seemed misguided.

'For the first time in history,' Carson charged, 'every
human being is now subjected to contact with dangerous
chemicals from the moment of conception until death.' Her
critique so impressed US President John F. Kennedy that
he ordered an examination of the subject of pesticide misuse by his President's Science Advisory Committee.

But Rachel Carson did more than challenge the scientific
establishment, or force the implementation of new pesticide
regulations. The hostile reaction of the establishment to
Carson and her book was evidence that many government
and industry officials recognized that Carson had not only
challenged the conclusions of scientists regarding the benefits of the new pesticides, but that she had undermined their
moral integrity and leadership. She had toppled America's
blind faith in science and, more damaging still, she initiated
public debate over the direction of technological progress.

In *Silent Spring* Carson illustrated the intricacies of ecology to a public that had only recently been forced to come
to grips with the reality of radioactive fallout. It is no accident that the first chemical Carson mentions in *Silent Spring*
is not DDT, but the radioactive element Strontium 90.
Carson wrote *Silent Spring* against a backdrop of secret
nuclear testing and stockpiling. She began her research
when the US military was trying to hide the details of the

atomic tests in the Bikini Islands, and published it a few months after the world was brought to the brink of nuclear holocaust with the Cuban Missile Crisis. She intended her message to protect and conserve the whole fabric of life, to convince humankind to act with humility rather than arrogance toward the rest of nature, and to see themselves as an integral part of it.

Carson was not the first writer in the post-war period to perceive the specific threat of pesticide abuse, or the first to speak out. But Rachel Carson was the voice the public listened to, and it was Carson's vision that gave shape to the ensuing social movement. She deliberately took on the most cherished tenets of the scientific establishment and, with an unquenchable anger at what she considered the 'senseless, brutish things' that human beings were doing in their war against nature, tried to make us look at what we were doing to life in the name of progress. She did so with no other motive than her own immense love of the living world, and with the romantic and perhaps naïve belief that if the public were made aware of the wonder and mystery of life, they would have less appetite to destroy it.

The corporate response to Carson's *Silent Spring* was initially cautious. The agricultural chemical industry and its allies in government treated the book simply as an annoying public relations problem. The chemical lobby had threatened to sue the *New Yorker*, Carson's publisher, and supportive conservation organizations to stop publication of the book. When this effort failed, the industry spent over a quarter of a million dollars in a publicity campaign designed to denigrate Carson's science. They tried to persuade the public that pesticides were beneficial, harmless and vital to the future of US agriculture, and that Rachel Carson's misguided conclusions would return civilization to the Dark Ages.

Their campaign only brought Carson and the message of *Silent Spring* more publicity. When the President's Science Advisory Committee issued their report on 5 May 1963, it vindicated Carson's evidence and concluded with the astonishing admission that 'until the publication of *Silent*

Spring, people were generally unaware of the toxicity of pesticides'. Frightened that their lucrative industry might be subjected to stringent government regulation, that millions of dollars in revenue might be lost by sagging public confidence in the safety and wisdom of using chemical pesticides, the agri-chemical industry and its allies decided to attack the messenger.

Rachel Carson was vulnerable to this well-financed and bitter personal attack for several reasons. First, she was a scientist without a doctorate and without institutional affiliation. She had no colleagues to defend her research, no protected forum from which to speak out against the hidden agendas of her detractors, and no organized network of support.

She was also a female working in public science at a time when women were not wanted in the academy and were not respected either as research scientists or social prophets. Added to this handicap of gender, Carson was encumbered by a financially and emotionally draining family, and after 1960, by an aggressively spreading, misdiagnosed breast cancer that manifested itself in a 'catalogue of illnesses' and which took her life barely sixteen months after *Silent Spring* was published.

Finally, and most detrimental to having her conclusions taken seriously, Carson was a scientist who wrote for the public; a calling the scientific establishment consistently denigrated. Critics cited her recognized literary achievements, three best-selling books, *Under the Sea-Wind*, *The Sea Around Us* and *The Edge of the Sea*, as if they proved she could not also be a real scientist. They inferred that her explanations of complex biology and chemistry were by definition inaccurate because they were too literate, too easily understood by the public.

Journalists and reviewers labelled Carson 'an hysterical woman', who used 'emotion-fanning words', a woman with an overly sensitive nature, whose book was 'more poisonous than the pesticides she condemns'.

The medical critic William B. Bean wrote, '*Silent Spring* . . . kept reminding me of trying to win an argument with a woman. It cannot be done.' She was an alarmist, they claimed, who resorted to unscientific fables, like the one that opens the book, to scare people. She kept cats and loved birds. She was a nature writer, a mystic, a romantic, sentimental woman who had taken on a subject that was beyond her intellectual ability. The chief spokesman for the chemical industry attacked Carson as 'a fanatic defender of the cult of the balance of nature', as though she were an unscientific practitioner of some branch of witchcraft. Even a former Secretary of Agriculture was known to wonder publicly 'why a spinster with no children was so interested in genetics'?

Behind these charges was understandable resentment of Carson's attack on the reputation and morality of the scientific establishment and, by inference, on a male-dominated technology. Carson's sin was not only that she had presented only one side of the argument, but her more unpardonable offence was that she had overstepped her place as a woman.

Carson was hurt and angered by these attacks. But instead of responding defensively, she calmly stood her ground and presented new and more incontrovertible evidence of pesticide damage and misuse at each public appearance. In a speech to the Women's National Press Club she took advantage of her independence and lack of affiliation, charging that basic scientific truths were being compromised to 'serve the gods of profit and production', and describing in detail the liaison between science and industry. 'When a scientific organization speaks,' Carson asked, 'whose voice do we hear, that of science or of the sustaining industry?'

Silent Spring and Rachel Carson could not be silenced. The book achieved enormous popularity and broad public support. Before Carson's death in April 1964 nearly a million copies had been sold. Carson looked nothing like an alarmist when she calmly and carefully answered questions on the American television news programme *CBS Reports*. A month later, before a subcommittee of the US Senate, her testimony in support of legislative reforms and more government-sponsored research set in motion a new spirit of activism. The first grass-roots environmental organizations appeared in many states, and legislation was introduced to limit the use of pesticides until more was known about its impact on wildlife and human health.

To the public for whom she had laboured to produce clear explanations of the unseen, but potentially damaging toxic substances that infiltrated air, water, soil and accumulated silently in the human body, there was no question that Rachel Carson's description of the intricacies of ecology were more right than wrong, and deserved further investigation. The same public that understood the workings of radioactive fallout recognized that the effects of pesticides on human health could be both immediate and long term; knew that they could include cancer, mutations and birth defects. For better or worse, concerns about human health as well as the fear of environmental destruction became the hallmarks of the modern environmental movement.

Carson and her warnings about chemical pollution have been embraced to the point that the ecological well-being of every community is considered essential to the common good. But at the beginning of the new millennium our world is awash with thousands of new and more damaging chemicals. Although DDT and PCBs have been banned in certain countries, the synergistic mix of pollutants has multiplied the dangers to all of nature and has left the earth a place of potentially unknown and undreamed of environmental horrors. Our legal, regulatory and political systems have failed so far to find a way to adequately clean up the earth and protect life in the future. The chemical industry, the agricultural community and the public health professionals in the industrial world have failed to find a meaningful way to communicate with each other. For all these reasons it is important to rediscover Rachel Carson and *Silent Spring*, and to take her warning and her hope to heart.

Silent Spring reminds us as almost no other piece of scientific literature not only of the practical workings of ecology, which we too easily ignore, but also of the power of the individual to bring about change. Rachel Carson's efforts to save what she loved left us a legacy of singular integrity and personal courage.

She understood the modern world as a troubled place. She deplored the increasing exploitation of resources and the cultural tendency to see the nature world as little more than an aggregate of impersonal commodities, rather than an integrated, organic and living whole.

Carson correctly feared the results of an unimpeded technology that initiates an action before fully knowing the consequences, and a culture that demands a quick fix for every problem. Such attitudes, she recognized, were formidable opponents to the cultivation of a sense of wonder and a reverence for the complex of intricate ecological relationships of the living world. In *Silent Spring* Carson gave these cultural qualities a historical context, protested their effects, warned against complacency, and offered a new ethic and a practical sort of hope. Hers is a message that we in the twenty-first century must find the courage to heed.

Linda Lear
Bethesda, Maryland
Spring 1998

List of Principal Sources

CHAPTER 2: THE OBLIGATION TO ENDURE

PAGE 24 'Report on Environmental Health Problems', *Hearings*, 86th Congress, Subcom. of Com. on Appropriations, March 1960, p. 170.

PAGE 26 *The Pesticide Situation for 1957-58*, U.S. Dept of Agric., Commodity Stabilization Service, April 1958, p. 10.

PAGE 27 Elton, Charles S., *The Ecology of Invasions by Animals and Plants*, New York, Wiley, 1958; London, Methuen, 1958.

PAGES 28-9 Shepard, Paul, 'The Place of Nature in Man's World', *Atlantic Naturalist*, Vol. 13 (April-June 1958), pp. 85-9.

CHAPTER 3: ELIXIRS OF DEATH

PAGES 31-49 Gleason, Marion, *et al.*, *Clinical Toxicology of Commercial Products*, Baltimore, Williams and Wilkins, 1957.

PAGES 31-49 Gleason, Marion, *et al.*, *Bulletin of Supplementary Material: Clinical Toxicology of Commercial Products*, Vol. IV, No. 9. Univ. of Rochester.

PAGE 32 *The Pesticide Situation for 1958-59*, U.S. Dept of Agric., Commodity Stabilization Service, April 1959, pp. 1-24

PAGE 32 *The Pesticide Situation for 1959-60*, U.S. Dept of Agric., Commodity Stabilization Service, July 1961; pp. 1-23.

PAGE 33 Hueper, W. C., *Occupational Tumors and Allied Diseases*, Springfield, Ill., Thomas, 1942.

PAGE 33 Todd, Frank E. and McGregor, S. E., 'Insecticides and Bees', *Yearbook of Agric.*, U.S. Dept of Agric., 1952, pp. 131-5.

PAGE 35 Bowen, C. V. and Hall, S. A., 'The Organic Insecticides', *Yearbook of Agric.*, U.S. Dept of Agric., 1952, pp. 209-18.

PAGE 36 Van Oettingen, W. F., *The Halogenated Aliphatic, Olefinic, Cyclic, Aromatic, and Aliphatic-Aromatic Hydrocarbons: Including the Halogenated Insecticides, Their Toxicity and Potential Dangers*. U.S. Dept of Health, Education, and Welfare. Public Health Service Publ. No. 414 (1955), pp. 341–2.

PAGE 36 Laug, Edwin P., *et al.*, 'Occurrence of DDT in Human Fat and Milk', *A.M.A. Archives Indus. Hygiene and Occupat. Med.*, Vol. 3 (1951), pp. 245–6.

PAGE 36 Biskind, Morton S., 'Public Health Aspects of the New Insecticides', *Am. Jour. Diges. Diseases*, Vol. 20 (1953), No. 11, pp. 331–41.

PAGE 36 Laug, Edwin P., *et al.*, 'Liver Cell Alteration and DDT Storage in the Fat of the Rat Induced by Dietary Levels of 1 to 50 p.p.m. DDT', *Jour. Pharmacol. and Exper. Therapeut.*, Vol. 98 (1950), p. 268.

PAGE 36 Ortega, Paul, *et al.*, 'Pathologic Changes in the Liver of Rats after Feeding Low Levels of Various Insecticides', *A.M.A. Archives Path.*, Vol. 64 (Dec. 1957), pp. 614–22.

PAGE 37 Fitzhugh, O. Garth and Nelson, A. A., 'The Chronic Oral Toxicity of DDT (2,2-BIS p-CHLOROPHENYL-1,1,1-TRI-CHLOROETHANE)', *Jour. Pharmacol. and Exper. Therapeut.*, Vol. 89 (1947), No. 1, pp. 18–30.

PAGE 37 Laug, Edwin P., *et al.*, 'Occurrence of DDT in Human Fat and Milk'.

PAGE 37 Hayes, Wayland, J., Jr, *et al.*, 'Storage of DDT and DDE in People with Different Degrees of Exposure to DDT', *A.M.A. Archives Indus. Health*, Vol. 18 (Nov. 1958), pp. 398–406.

PAGE 37 Durham, William F., *et al.*, 'Insecticide Content of Diet and Body Fat of Alaskan Natives', *Science*, Vol. 134 (1961), No. 3493, pp. 1880–1.

PAGE 37 Van Oettingen, W. F., *Halogenated . . . Hydrocarbons*, p. 363.

PAGE 37 Smith, Ray F., *et al.*, 'Secretion of DDT in Milk of Dairy Cows Fed Low Residue Alfalfa', *Jour. Econ. Entomol.*, Vol. 41 (1948), pp. 759–63.

PAGES 37–8 Laug, Edwin P., *et al.*, 'Occurrence of DDT in Human Fat and Milk'.

PAGE 38 Finnegan, J. K., *et al.*, 'Tissue Distribution and Elimination of DDD and DDT Following Oral Administration to Dogs and Rats', *Proc. Soc. Exper. Biol and Med.*, Vol. 72 (1949), 356–7.

PAGE 38 Laug, Edwin P., *et al.*, 'Liver Cell Alteration'.

PAGE 38 'Chemicals in Food Products', *Hearings*, H.R. 74, House Select Com. to Investigate Use of Chemicals in Food Products, Pt 1 (1951), p. 275.

PAGE 38 Van Oettingen, W. F., *Halogenated . . . Hydrocarbons*, p. 322.

PAGE 39 'Chemicals in Food Products', *Hearings*, 81st Congress, H.R. 323, Com. to Investigate Use of Chemicals in Food Products, Pt 1 (1950), pp. 388–90.

PAGE 39 *Clinical Memoranda on Economic Poisons.* U.S. Public Health Service Publ. No. 476 (1956), p. 28.

PAGE 39 Gannon, Norman and Bigger, J. H., 'The Conversion of Aldrin and Heptachlor to Their Epoxides in Soil', *Jour. Econ. Entomol.*, Vol. 51 (February 1958), pp. 1–2.

PAGE 39 Davidow, B. and Radomski, J. L., 'Isolation of an Epoxide Metabolite from Fat Tissues of Dogs Fed Heptachlor', *Jour. Pharmacol. and Exper. Therapeut.*, Vol. 107 (March 1953), pp. 259–65.

PAGE 39 Van Oettingen, W. F., *Halogenated . . . Hydrocarbons*, p. 310.

PAGE 39 Drinker, Cecil K., *et al.*, 'The Problem of Possible Systemic Effects from Certain Chlorinated Hydrocarbons', *Jour. Indus. Hygiene and Toxicol.*, Vol. 19 (September 1937), p. 283.

PAGE 39 'Occupational Dieldrin Poisoning', Com. on Toxicology, *Jour. Am. Med. Assn.*, Vol. 172 (April 1960), pp. 2077–80.

PAGE 39 Scott, Thomas G., *et al.*, 'Some Effects of a Field Application of Dieldrin on Wildlife', *Jour. Wildlife Management*, Vol. 23 (October 1959), pp. 409–27.

PAGE 39 Paul, A. H., 'Dieldrin Poisoning – A Case Report', *New Zealand Med. Jour.*, Vol. 58 (1959), p. 393.

PAGE 40 Hayes, Wayland J., Jr, 'The Toxicity of Dieldrin to Man', *Bull. World Health Organ.*, Vol. 20 (1959), pp. 891–912.

PAGE 40   Gannon, Norman and Decker, G. C., 'The Conversion of Aldrin to Dieldrin on Plants', *Jour. Econ. Entomol.*, Vol. 51 (February 1958), pp. 8–11.

PAGE 40   Kitselman, C. H., *et al.*, 'Toxicological Studies of Aldrin (Compound 118) on Large Animals', *Am. Jour. Vet. Research*, Vol. 11 (1950), p. 378.

PAGE 40   Dahlen, James H. and Haugen, A. O., 'Effect of Insecticides on Quail and Doves', *Alabama Conservation*, Vol. 26 (1954), No. 1, pp. 21–3.

PAGE 40   De Witt, James B., 'Chronic Toxicity to Quail and Pheasants of Some Chlorinated Insecticides', *Jour. Agric. and Food Chem.*, Vol. 4 (1956), No. 10, pp. 863–6.

PAGE 40   Kitselman, C. H., 'Long Term Studies on Dogs Fed Aldrin and Dieldrin in Sublethal Doses, with Reference to the Histopathological Findings and Reproduction', *Jour. Am. Vet. Med. Assn.*, Vol. 123 (1953), p. 28.

PAGE 40   Treon, J. F. and Borgmann, A. R., 'The Effects of the Complete Withdrawal of Food from Rats Previously Fed Diets Containing Aldrin or Dieldrin', Kettering Lab., Univ. of Cincinnati; mimeo. Quoted from Robert L. Rudd and Richard E. Genelly, *Pesticides: Their Use and Toxicity in Relation to Wildlife*. Calif. Dept of Fish and Game, Game Bulletin No. 7 (1956), p. 52.

PAGES 40–1   Myers, C. S., 'Endrin and Related Pesticides: A Review', Pennsylvania Dept of Health Research Report No. 45 (1958). Mimeo.

PAGES 40–1   Jacobziner, Harold and Raybin, H. W., 'Poisoning by Insecticide (Endrin)', *New York State Jour. Med.*, Vol. 59 (15 May 1959), pp. 2017–22.

PAGE 41   'Care in Using Pesticide Urged', *Clean Streams*, No. 46 (June 1959). Pennsylvania Dept of Health.

PAGE 41   Metcalf, Robert L., 'The Impact of the Development of Organophosphorus Insecticides upon Basic and Applied Science', *Bull. Entomol. Soc. Am.*, Vol. 5 (March 1959), pp. 3–15.

PAGES 42–3   Mitchell, Philip H., *General Physiology*, New York, McGraw-Hill, 1958, pp. 14–15.

PAGE 43   Brown, A. W. A., *Insect Control by Chemicals*. New York, Wiley, 1951; London, Chapman & Hall, 1951.

PAGE 43   Toivonen, T., *et al.*, 'Parathion Poisoning Increasing Frequency in Finland', *Lancet*, Vol. 2 (1959), No. 7095, pp. 175–6.

PAGES 43–4   Hayes, Wayland J., Jr, 'Pesticides in Relation to Public Health', *Annual Rev. Entomol.*, Vol. 5 (1960), pp. 379–404.

PAGES 43–4   Quinby, Griffith E. and Lemmon, A. B., 'Parathion Residues As a Cause of Poisoning in Crop Workers', *Jour. Am. Med. Assn.*, Vol. 166 (15 February 1958), pp. 740–6.

PAGES 43–4   Carman, G. C., *et al.*, 'Absorption of DDT and Parathion by Fruits', *Abstracts*, 115th Meeting Am. Chem. Soc. (1948), p. 30A.

PAGE 44   *Clinical Memoranda on Economic Poisons*, p. 11.

PAGE 44   *Occupational Disease in California Attributed to Pesticides and Other Agricultural Chemicals.* California Dept of Public Health, 1957, 1958, 1959, and 1960.

PAGE 45   Frawley, John P., *et al.*, 'Marked Potentiation in Mammalian Toxicity from Simultaneous Administration of Two Anticholinesterase Compounds', *Jour. Pharmacol. and Exper. Therapeut.*, Vol. 121 (1957), No. 1, pp. 96–106.

PAGE 45   Rosenberg, Phillip and Coon, J. M., 'Potentiation between Cholinesterase Inhibitors', *Proc. Soc. Exper. Biol. and Med.*, Vol. 97 (1958), pp. 836–9.

PAGE 45   Dubois, Kenneth P., 'Potentiation of the Toxicity of Insecticidal Organic Phosphates', *A.M.A. Archives Indus. Health*, Vol. 18 (December 1958), pp. 488–96.

PAGE 45   Murphy, S. D., *et al.*, 'Potentiation of Toxicity of Malathion by Triorthotolyl Phosphate', *Proc. Soc. Exper. Biol. and Med.*, Vol. 100 (March 1959), pp. 483–7.

PAGE 45   Graham, R. C. B., *et al.*, 'The Effect of Some Organophosphorus and Chlorinated Hydrocarbon Insecticides on the Toxicity of Several Muscle Relaxants', *Jour. Pharm. and Pharmacol.*, Vol. 9 (1957), pp. 312–19.

PAGE 45   Rosenberg, Philip and Coon, J. M., 'Increase of Hexobarbital Sleeping Time by Certain Anticholinesterases', *Proc. Soc. Exper. Biol. and Med.*, Vol. 98 (1958), pp. 650–2.

PAGE 46   Dubois, Kenneth P., 'Potentiation of Toxicity'.

PAGE 46   Hurd-Karrer, A. M. and Poos, F. W., 'Toxicity of

Selenium-Containing Plants to Aphids', *Science*, Vol. 84 (1936), pp. 252.

PAGE 46 Ripper, W. E., 'The Status of Systemic Insecticides in Pest Control Practices', *Advances in Pest Control Research*, New York, Interscience, 1957. Vol. I, pp. 305–52.

PAGES 46–7 *Occupational Disease in California*, 1959.

PAGE 47 Glynne-Jones, G. D. and Thomas, W. D. E., 'Experiments on the Possible Contamination of Honey with Schradan', *Annals Appl. Biol.*, Vol. 40 (1953), p. 546.

PAGE 47 Radeleff, R. D., *et al.*, *The Acute Toxicity of Chlorinated Hydrocarbon and Organic Phosphorus Insecticides to Livestock*, U.S. Dept of Agric. Technical Bulletin 1122 (1955).

PAGE 48 Brooks, F. A., 'The Drifting of Poisonous Dusts Applied by Airplanes and Land Rigs', *Agric. Engin.*, Vol. 28 (1947), No. 6, pp. 233–9.

PAGE 48 Stevens, Donald B., 'Recent Developments in New York State's Program Regarding Use of Chemicals to Control Aquatic Vegetation', paper presented at 13th Annual Meeting North-eastern Weed Control Conf. (8 January 1959).

PAGE 48 Anon., 'No More Arsenic', *Economist*, 10 October 1959.

PAGES 48–9 'Arsenites in Agriculture', *Lancet*, Vol. I (1960), p. 178.

PAGES 48–9 Horner, Warren D., 'Dinitrophenol and Its Relation to Formation of Cataract', (A.M.A.) *Archives Ophthalmol.*, Vol. 27 (1942), pp. 1097–121.

PAGE 49 Weinbach, Eugene C., 'Biochemical Basis for the Toxicity of Pentachlorophenol', *Proc. Natl. Acad. Sci.*, Vol. 43 (1957), No. 5, pp. 393–7.

CHAPTER 4: SURFACE WATERS AND UNDERGROUND SEAS

PAGE 50 *Biological Problems in Water Pollution*. Transactions, 1959 seminar. U.S. Public Health Service Technical Report W60–3 (1960).

PAGE 51 'Report on Environmental Health Problems', *Hearings*, 86th Congress, Subcom. of Com. on Appropriations, March 1960, p. 78.

PAGE 51   Tarzwell, Clarence M., 'Pollutional Effects of Organic Insecticides to Fishes', *Transactions*, 24th North Am. Wildlife Conf. (1959), Washington, D.C., pp. 132–42. Published by Wildlife Management Inst.

PAGE 52   Nicholson, H. Page, 'Insecticide Pollution of Water Resources', *Jour. Am. Waterworks Assn.*, Vol. 51 (1959), pp. 981–6.

PAGE 52   Woodward, Richard L., 'Effects of Pesticides in Water Supplies', *Jour. Am. Waterworks Assn.*, Vol. 52 (1960), No. 11, pp. 1367–72.

PAGE 52   Cope, Oliver B., 'The Retention of DDT by Trout and Whitefish', in *Biological Problems in Water Pollution*, pp. 72–5.

PAGE 53   Kuenen, P. H., *Realms of Water*, New York, Wiley, 1955; London, Cleaver-Hume Press, 1955.

PAGE 53   Gilluly, James, *et al.*, *Principles of Geology*, San Francisco, Freeman, 1951.

PAGES 53–4   Walton, Graham, 'Public Health Aspects of the Contamination of Ground Water in South Platte River Basin in Vicinity of Henderson, Colorado, August, 1959'. U.S. Public Health Service, 2 November 1959. Mimeo.

PAGES 53–4   'Report on Environmental Health Problems'.

PAGE 55   Hueper, W. C., 'Cancer Hazards from Natural and Artificial Water Pollutants', *Proc.*, Conf. on Physiol. Aspects of Water Quality, Washington, D.C., 8–9 September 1960. U.S. Public Health Service.

PAGES 55–9   Hunt, E. G. and Bischoff, A. I., 'Inimical Effects on Wildlife of Periodic DDD Applications to Clear Lake', *Calif. Fish and Game*, Vol. 46 (1960), No. 1, pp. 91–106.

PAGE 59   Woodward, G., *et al.*, 'Effects Observed in Dogs Following the Prolonged Feeding of DDT and Its Analogues', *Federation Proc.*, Vol. 7 (1948), No. 1, p. 266.

PAGE 59   Nelson, A. A. and Woodward, G., 'Severe Adrenal Cortical Atrophy (Cytotoxic) and Hepatic Damage Produced in Dogs by Feeding DDD or TDE', (A.M.A.) *Archives Path.*, Vol. 48 (1949), p. 387.

PAGES 58–9   Zimmermann, B., *et al.*, 'The Effects of DDD on the Human Adrenal; Attempts to Use an Adrenal-Destructive Agent in the Treatment of Disseminated Mammary and Prostatic Cancer', *Cancer*, Vol. 9 (1956), pp. 940–8.

PAGES 59–60   Cohen, Jesse M., *et al.*, 'Effect of Fish Poisons on Water Supplies. I. Removal of Toxic Materials', *Jour. Am. Waterworks Assn.*, Vol. 52 (1960), No. 12, pp. 1551–65. 'II. Odor Problems', Vol. 53 (1960), No. 1, pp. 49–61. 'III. Field Study, Dickinson, North Dakota', Vol. 53 (1961), No. 2, pp. 233–46.

PAGE 60   Hueper, W. C., 'Cancer Hazards from Water Pollutants'.

CHAPTER 5: REALMS OF THE SOIL

PAGE 61   Simonson, Roy W., 'What Soils Are', *Yearbook of Agric.*, U.S. Dept of Agric., 1957, pp. 17–31.

PAGE 62   Clark, Francis E., 'Living Organisms in the Soil', *Yearbook of Agric.*, U.S. Dept of Agric., 1957, pp. 157–65.

PAGE 63   Farb, Peter, *Living Earth*, New York, Harper, 1959; London, Constable, 1960.

PAGE 64   Lichtenstein, E. P. and Schulz, K. R., 'Persistence of Some Chlorinated Hydrocarbon Insecticides As Influenced by Soil Types, Rate of Application and Temperature', *Jour. Econ. Entomol.*, Vol. 52 (1959), No. 1, pp. 124–31.

PAGE 64   Thomas, F. J. D., 'The Residual Effects of Crop-Protection Chemicals in the Soil', in *Proc.*, 2nd Internatl Plant Protection Conf. (1956), Fernhurst Research Station, England.

PAGE 64   Eno, Charles, F., 'Chlorinated Hydrocarbon Insecticides: What Have They Done to Our Soil?' *Sunshine State Agric. Research Report* for July 1959.

PAGE 64   Mader, Donald L., 'Effect of Humus of Different Origin in Moderating the Toxicity of Biocides'. Doctorate thesis, Univ. of Wisc., 1960.

PAGES 64–5   Cullinan, F. P., 'Some New Insecticides – Their Effect on Plants and Soils', *Jour. Econ. Entomol.*, Vol. 42 (1949), pp. 387–91.

PAGE 65   Sheals, J. G., 'Soil Population Studies. I. The Effects of Cultivation and Treatment with Insecticides', *Bull. Entomol. Research*, Vol. 47 (December 1956), pp. 803–22.

PAGE 65   Hetrick, L. A., 'Ten Years of Testing Organic Insecticides As Soil Poisons against the Eastern Subterranean Termite', *Jour. Econ. Entomol.*, Vol. 50 (1957), p. 316.

PAGE 65 Lichtenstein, E. P. and Polivka, J. B., 'Persistence of Insecticides in Turf Soils', *Jour. Econ. Entomol.*, Vol. 52 (1959), No. 2, pp. 289–93.

PAGE 65 Ginsburg, J. M. and Reed, J. P., 'A Survey on DDT-Accumulation in Soils in Relation to Different Crops', *Jour. Econ. Entomol.*, Vol. 47 (1954), No. 3, pp. 467–73.

PAGE 66 Satterlee, Henry S., 'The Problem of Arsenic in American Cigarette Tobacco', *New Eng. Jour. Med.*, Vol. 254 (21 June 1956), pp. 1149–54.

PAGE 66 Lichtenstein, E. P., 'Absorption of Some Chlorinated Hydrocarbon Insecticides from Soils into Various Crops', *Jour. Agric. and Food Chem.*, Vol. 7 (1959), No. 6, pp. 430–3.

PAGES 66–7 'Chemicals in Foods and Cosmetics', *Hearings*, 81st Congress, H.R. 74 and 447, House Select Com. to Investigate Use of Chemicals in Foods and Cosmetics, Pt 3 (1952), pp. 1385–416. Testimony of L. G. Cox.

PAGES 67–8 Klostermeyer, E. C. and Skotland, C. B., *Pesticide Chemicals As a Factor in Hop Die-out*. Washington Agric. Exper. Stations Circular 362 (1959).

PAGE 68 Stegeman, LeRoy C., 'The Ecology of the Soil'. Transcription of a seminar, New York State Univ. College of Forestry, 1960.

CHAPTER 6: EARTH'S GREEN MANTLE

PAGES 69–71 Patterson, Robert L., *The Sage Grouse in Wyoming*, Denver, Sage Books, for Wyoming Fish and Game Commission, 1952.

PAGES 70–1 Murie, Olaus J., 'The Scientist and Sagebrush', *Pacific Discovery*, Vol. 13 (1960), No. 4, p. 1.

PAGE 71 Pechanec, Joseph, *et al.*, *Controlling Sagebrush on Rangelands*. U.S. Dept of Agric. Farmers' Bulletin No. 2072 (1960).

PAGES 72–3 Douglas, William O., *My Wilderness: East to Katahdin*, New York, Doubleday, 1961.

PAGE 73 Egler, Frank E., *Herbicides: 60 Questions and Answers Concerning Roadside and Rightofway Vegetation Management*, Litchfield, Conn., Litchfield Hills Audubon Soc., 1961.

PAGE 73 Fisher, C. E., *et al.*, *Control of Mesquite on Grazing*

Lands. Texas Agric. Exper. Station Bulletin 935 (August 1959).

PAGE 73   Goodrum, Phil D. and Reid, V. H., 'Wildlife Implications of Hardwood and Brush Controls', *Transactions*, 21st North Am. Wildlife Conf. (1956).

PAGE 73   *A Survey of Extent and Cost of Weed Control and Specific Weed Problems*. U.S. Dept of Agric. ARS 34–23 (March 1962).

PAGE 74   Barnes, Irston R., 'Sprays Mar Beauty of Nature', *Washington Post*, 25 September 1960.

PAGE 75   Goodwin, Richard H. and Niering, William A, *A Roadside Crisis: The Use and Abuse of Herbicides*. Connecticut Arboretum Bulletin No. 11 (March 1959), pp. 1–13.

PAGE 75   Boardman, William, 'The Dangers of Weed Spraying', *Veterinarian*, Vol. 6 (January 1961), pp. 9–19.

PAGE 76   Willard, C. J., 'Indirect Effects of Herbicides', *Proc.*, 7th Annual Meeting North Central Weed Control Conf. (1950), pp. 110–12.

PAGE 76   Douglas, William O., *My Wilderness: The Pacific West*, New York, Doubleday, 1960.

PAGE 77   Egler, Frank E., *Vegetation Management for Rights-of-Way and Roadsides*. Smithsonian Report for 1953 (Smithsonian Inst., Washington, D.C.), pp. 299–322.

PAGE 77   Bohart, George E., 'Pollination by Native Insects', *Yearbook of Agric.*, U.S. Dept of Agric., 1952, pp. 107–21.

PAGE 78   Egler, *Vegetation Management*.

PAGES 78–9   Niering, William A. and Egler, Frank E., 'A Shrub Community of *Viburnum lentago*, Stable for Twenty-five Years', *Ecology*, Vol. 36 (April 1955), pp. 356–60.

PAGE 79   Pound Charles E. and Egler, Frank E., 'Brush Control in South-eastern New York: Fifteen Years of Stable Tree-less Communities', *Ecology*, Vol. 34 (January 1953), pp. 63–73.

PAGE 79   Egler, Frank E., 'Science, Industry and the Abuse of Rights of Way', *Science*, Vol. 127 (1958), No. 3298, pp. 573–80.

PAGE 79   Niering, William A., 'Principles of Sound Right-of-Way Vegetation Management', *Econ. Botany*, Vol. 12 (April–June 1958), pp. 140–4.

PAGE 79   Hall, William C. and Niering, William A., 'The Theory and Practice of Successful Selective Control of "Brush" by Chemicals', *Proc.*, 13th Annual Meeting North-eastern Weed Control Conf. (8 January 1959).

PAGE 79   Egler, Frank E., 'Fifty Million More Acres for Hunting?' *Sports Afield*, December 1954.

PAGE 79   McQuilkin; W. E. and Strickenberg, L. R., *Roadside Brush Control with 2,4,5-T on Eastern National Forests*. North-eastern Forest Exper. Station Paper No. 148. Upper Darby, Pennsylvania, 1961.

PAGE 80   Goldstein, N. P., *et al.*, 'Peripheral Neuropathy after Exposure to an Ester of Dichlorophenoxyacetic Acid', *Jour. Am. Med. Assn.*, Vol. 171 (1959), pp. 1306–9.

PAGE 80   Brody, T. M., 'Effect of Certain Plant Growth Sub-stances on Oxidative Phosphorylation in Rat Liver Mito-chondria', *Proc. Soc. Exper. Biol. and Med.*, Vol. 80 (1952), pp. 533–6.

PAGE 80   Croker, Barbara H., 'Effects of 2,4-D and 2,4,5-T on Mitosis in *Allium cepa*', *Bot. Gazette*, Vol. 114 (1953), pp. 274–83.

PAGE 80   Willard, C. J., 'Indirect Effects of Herbicides'.

PAGES 80–1   Stahler, L. M., and Whitehead, E. J., 'The Effect of 2,4-D on Potassium Nitrate Levels in Leaves of Sugar Beets', *Science*, Vol. 112 (1950), No. 2921, pp. 749–51.

PAGES 80–1   Olson, O. and Whitehead,. E. 'Nitrate Content of Some South Dakota Plants', *Proc.*, South Dakota Acad. of Sci., Vol. 20 (1940), p. 95.

PAGES 80–1   Stahler, L. M. and Whitehead, E. J., 'The Effect of 2,4-D on Potassium Nitrate Levels'.

PAGE 81   *What's New in Farm Science*. Univ. of Wisc. Agric. Exper. Station Annual Report, Pt II, Bulletin 527 (July 1957), p. 18.

PAGE 81   Grayson, R. R., 'Silage Gas Poisoning: Nitrogen Dioxide Pneumonia, a New Disease in Agricultural Workers', *Annals Internal Med.*, Vol. 45 (1956), pp. 393–408.

PAGE 81   Crawford, R. F. and Kennedy, W. K., *Nitrates in Forage Crops and Silage: Benefits, Hazards, Precautions*. New York State College of Agric., Cornell Misc. Bulletin 37 (June 1960).

PAGE 82 Briejèr, C. J., To author.

PAGE 83 Knake, Ellery L. and Slife, F. W., 'Competition of *Setaria faterii* with Corn and Soybeans', *Weeds*, Vol. 10 (1962), No. 1, pp. 26–9.

PAGE 83 Goodwin, Richard H. and Niering, William A., *A Roadside Crisis*.

PAGE 83 Egler, Frank E., To author.

PAGE 83 DeWitt, James B., To author.

PAGE 83 Holloway, James K., 'Weed Control by Insect', *Sci. American*, Vol. 197 (1957), No. 1, pp. 56–62.

PAGE 83 Holloway, James K. and Huffaker, C. B., 'Insects to Control a Weed', *Yearbook of Agric.*, U.S. Dept of Agric., 1952, pp. 135–40.

PAGE 83 Huffaker, C. B. and Kennett, C. E., 'A Ten-Year Study of Vegetational Changes Associated with Biological Control of Klamath Weed', *Jour. Range Management*, Vol. 12 (1959), No. 2, pp. 69–82.

PAGE 86 Bishopp, F. C., 'Insect Friends of Man', *Yearbook of Agric.*, U.S. Dept of Agric., 1952, pp. 79–87.

CHAPTER 7: NEEDLESS HAVOC

PAGES 88–9 *Here Is Your 1959 Japanese Beetle Control Program*. Release, Michigan State Dept of Agric., 19 October 1959.

PAGE 89 Nickell, Walter, To author.

PAGE 89 Hadley, Charles H. and Fleming, Walter E., 'The Japanese Beetle', *Yearbook of Agric.*, U.S. Dept of Agric., 1952, pp. 567–73.

PAGE 90 *Here Is Your 1959 Japanese Beetle Control Program*.

PAGE 90 'No Bugs in Plane Dusting', *Detroit News*, 10 November 1959.

PAGE 91 *Michigan Audubon Newsletter*, Vol. 9 (January 1960).

PAGE 92 'No Bugs in Plane Dusting'.

PAGE 92 Hickey, Joseph J., 'Some Effects of Insecticides on Terrestrial Birdlife', *Report* of Subcom. on Relation of Chemicals to Forestry and Wildlife, Madison, Wisc., January 1961. Special Report No. 6.

PAGE 93 Scott, Thomas G., To author, 14 December 1961.

PAGE 93 'Coordination of Pesticides Programs', *Hearings*, 86th Congress, H.R. 11502, Com. on Merchant Marine and Fisheries, May 1960, p. 66.

PAGES 92-5 Scott, Thomas G., *et al.*, 'Some Effects of a Field Application of Dieldrin on Wildlife', *Jour. Wildlife Management*, Vol. 23 (1959), No. 4, pp. 409-27.

PAGE 94 Hayes, Wayland J., Jr, 'The Toxicity of Dieldrin to Man', *Bull. World Health Organ.*, Vol. 20 (1959), pp. 891-912.

PAGE 95 Scott, Thomas G., To author, 14 December 1961; 8 January, 15 February 1962.

PAGES 96-9 Hawley, Ira M., 'Milky Diseases of Beetles', *Yearbook of Agric.*, U.S. Dept of Agric., 1952, pp. 394-401.

PAGES 96-9 Fleming, Walter E., 'Biological Control of the Japanese Beetle Especially with Entomogenous Diseases', *Proc.*, 10th Internatl Congress of Entomologists (1956), Vol. 3 (1958), pp. 115-25.

PAGE 98 Chittick, Howard A. (Fairfax Biological Lab.), To author, 30 November 1960.

PAGE 99 Scott, Thomas G., *et al.*, 'Some Effects of a Field Application of Dieldrin on Wildlife'.

CHAPTER 8: AND NO BIRDS SING

PAGE 101 *Audubon Field Notes*. 'Fall Migration – Aug. 16 to Nov. 30, 1958', Vol. 13 (1959), No. 1, pp. 1-68.

PAGE 102 Swingle, R. U., *et al.*, 'Dutch Elm Disease', *Yearbook of Agric.*, U.S. Dept of Agric., 1949, pp. 451-2.

PAGE 102 Mehner, John F. and Wallace, George J., 'Robin Populations and Insecticides', *Atlantic Naturalist*, Vol. 14 (1959), No. 1, pp. 4-10.

PAGE 103 Wallace, George J., 'Insecticides and Birds', *Audubon Mag.*, January–February 1959.

PAGES 103-4 Barker, Roy J., 'Notes on Some Ecological Effects of DDT Sprayed on Elms', *Jour. Wildlife Management*, Vol. 22 (1958), No. 3, pp. 269-74.

PAGES 103-4 Hickey, Joseph J. and Hunt, L. Barrie, 'Songbird Mortality Following Annual Programs to Control

Dutch Elm Disease', *Atlantic Naturalist*, Vol. 15 (1960), No. 2, pp. 87–92.

PAGES 104–5   Wallace, George J., 'Insecticides and Birds'.

PAGES 104–5   Wallace, George J., 'Another Year of Robin Losses on a University Campus', *Audubon Mag.*, March–April 1960.

PAGE 105   'Coordination of Pesticides Programs', *Hearings*, H.R. 11502, 86th Congress, Com. on Merchant Marine and Fisheries, May 1960, pp. 10, 12.

PAGE 105   Hickey, Joseph J. and Hunt, L. Barrie, 'Initial Songbird Mortality Following a Dutch Elm Disease Control Program', *Jour. Wildlife Management*, Vol. 24 (1960), No. 3, pp. 259–65.

PAGE 105   Wallace, George J., *et al.*, *Bird Mortality in the Dutch Elm Disease Program in Michigan.* Cranbrook Inst. of Science Bulletin 41 (1961).

PAGE 105   Hickey, Joseph J., 'Some Effects of Insecticides on Terrestrial Birdlife', *Report* of Subcom. on Relation of Chemicals to Forestry and Wildlife, State of Wisconsin, January 1961, pp. 2–43.

PAGES 105–6   Wallace, George J., *et al.*, *Bird Mortality in the Dutch Elm Disease Program.*

PAGE 106   Walton, W. R., *Earthworms As Pests and Otherwise*, U.S. Dept of Agric. Farmers' Bulletin No. 1569 (1928).

PAGE 106   Wright, Bruce S., 'Woodcock Reproduction in DDT-Sprayed Areas of New Brunswick', *Jour. Wildlife Management*, Vol. 24 (1960), No. 4, pp. 419–20.

PAGE 106   Dexter, R. W., 'Earthworms in the Winter Diet of the Opossum and the Raccoon', *Jour. Mammal.*, Vol. 32 (1951), p. 464.

PAGE 107   'Coordination of Pesticides Programs'. Testimony of George J. Wallace, p. 10.

PAGES 107–8   Wallace, George J., 'Insecticides and Birds'.

PAGE 108   Bent, Arthur C., *Life Histories of North American Jays, Crows, and Titmice.* Smithsonian Inst., U.S. Natl Museum Bulletin 191 (1946).

PAGE 108   MacLellan, C. R., 'Woodpecker Control of the Codling Moth in Nova Scotia Orchards', *Atlantic Naturalist*, Vol. 16 (1961), No. 1, pp. 17–25.

PAGE 108 Knight, F. B., 'The Effects of Woodpeckers on Populations of the Engelmann Spruce Beetle', *Jour. Econ. Entomol.*, Vol. 51 (1958), pp. 603–7.

PAGE 109 Carter, J. C., To author, 16 June 1960.

PAGE 110 Sweeney, Joseph A., To author, 7 March 1960.

PAGE 110 Welch, D. S. and Matthysse, J. G., *Control of the Dutch Elm Disease in New York State*. New York State College of Agric., Cornell Ext. Bulletin No. 932 (June 1960), pp. 3–16.

PAGE 111 Miller, Howard, To author, 17 January 1962.

PAGE 112 Matthysse, J. G., *An Evaluation of Mist Blowing and Sanitation in Dutch Elm Disease Control Programs*. New York State College of Agric., Cornell Ext. Bulletin No. 30 (July 1959), pp. 2–16.

PAGE 112 Elton, Charles S., *The Ecology of Invasions by Animals and Plants*, New York, Wiley, 1958; London, Methuen, 1958.

PAGE 113 Broley, Charles E., 'The Bald Eagle in Florida', *Atlantic Naturalist*, July 1957, pp. 230–1.

PAGE 113 Broley, Charles E., 'The Plight of the American Bald Eagle', *Audubon Mag.*, July–August 1958, pp. 162–3.

PAGE 113 McLaughlin, Frank, 'Bald Eagle Survey in New Jersey', *New Jersey Nature News*, Vol. 16 (1959), No. 2, p. 25. Interim Report, Vol. 16 (1959), No. 3, p. 51.

PAGES 113–14 Cunningham, Richard L., 'The Status of the Bald Eagle in Florida', *Audubon Mag.*, January–February 1960, pp. 24–43.

PAGE 114 'Vanishing Bald Eagle Gets Champion', *Florida Naturalist*, April 1959, p. 64.

PAGE 114 Broun, Maurice, To author, 22, 30 May 1960.

PAGES 114–15 Beck, Herbert H., To author, 30 July 1959.

PAGE 115 De Witt, James B., 'Effects of Chlorinated Hydrocarbon Insecticides upon Quail and Pheasants', *Jour. Agric. and Food Chem.*, Vol. 3 (1955), No. 8, p. 672.

PAGE 115 De Witt, James B., 'Chronic Toxicity to Quail and Pheasants of Some Chlorinated Insecticides', *Jour. Agric. and Food Chem.*, Vol. 4 (1956), No. 10, p. 863.

PAGE 115 Rudd, Robert L. and Genelly, Richard E., *Pesticides: Their Use and Toxicity in Relation to Wildlife*. Calif. Dept of Fish and Game, Game Bulletin No. 7 (1956), p. 57.

PAGE 116   Imler, Ralph H. and Kalmbach, E. R., *The Bald Eagle and Its Economic Status*. U.S. Fish and Wildlife Service Circular 30 (1955).

PAGE 116   Mills, Herbert R., 'Death in the Florida Marshes', *Audubon Mag.*, September–October 1952.

PAGE 117   *Bulletin*, Internatl Union for the Conservation of Nature, May and October 1957.

PAGE 117   *The Deaths of Birds and Mammals Connected with Toxic Chemicals in the First Half of 1960*. Report No. 1 of the British Trust for Ornithology and Royal Soc. for the Protection of Birds. Com. on Toxic Chemicals, Royal Soc. Protect. Birds.

PAGES 117–19   *Sixth Report* from the Estimates Com., Ministry of Agric., Fisheries and Food, Sess. 1960–1, House of Commons.

PAGE 118   Christian, Garth, 'Do Seed Dressings Kill Foxes?' *Country Life*, 12 January 1961.

PAGE 119   Rudd, Robert L. and Genelly, Richard E., 'Avian Mortality from DDT in Californian Rice Fields', *Condor*, Vol. 57 (March–April 1955), pp. 117–18.

PAGES 119–20   Rudd, Robert L. and Genelly, Richard E., *Pesticides*.

PAGE 120   Dykstra, Walter W., 'Nuisance Bird Control', *Audubon Mag.*, May–June 1960, pp. 118–19.

PAGE 120   Buchheister, Carl W., 'What About Problem Birds?' *Audubon Mag.*, May–June 1960, pp. 116–18.

PAGE 120   Quinby, Griffith E. and Lemmon, A. B., 'Parathion Residues As a Cause of Poisoning in Crop Workers', *Jour. Am. Med. Assn.*, Vol. 166 (15 February 1958), pp. 740–6.

CHAPTER 9: RIVERS OF DEATH

PAGES 122–6   Kerswill, C. J., 'Effects of DDT Spraying in New Brunswick on Future Runs of Adult Salmon', *Atlantic Advocate*, Vol. 48 (1958), pp. 65–8.

PAGES 122–6   Keenleyside, M. H. A., 'Insecticides and Wildlife', *Canadian Audubon*, Vol. 21 (1959), No. 1, pp. 1–7.

PAGES 122–6   Keenleyside, M. H. A., 'Effects of Spruce Budworm Control on Salmon and Other Fishes in New Brunswick', *Canadian Fish Culturist*, Issue 24 (1959), pp. 17–22.

PAGES 122–6 Kerswill, C. J., *Investigation and Management of Atlantic Salmon in 1956* (also for 1957, 1958, 1959–60; in 4 parts). Federal-Provincial Co-ordinating Com. on Atlantic Salmon (Canada).

PAGE 124 Ide, F. P., 'Effect of Forest Spraying with DDT on Aquatic Insects of Salmon Streams', *Transactions*, Am. Fisheries Soc., Vol. 86 (1957), pp. 208–19.

PAGE 125 Kerswill, C. J., To author, 9 May 1961.

PAGES 125–6 Kerswill, C. J., To author, 1 June 1961.

PAGES 126–7 Warner, Kendall and Fenderson, O. C., 'Effects of Forest Insect Spraying on Northern Maine Trout Streams'. Maine Dept of Inland Fisheries and Game, Mimeo., n.d.

PAGE 127 Alderdice, D. F. and Worthington, M. E., 'Toxicity of a DDT Forest Spray to Young Salmon', *Canadian Fish Culturist*, Issue 24 (1959), pp. 41–8.

PAGE 127 Hourston, W. R., To author, 23 May 1961.

PAGE 128 Graham, R. J. and Scott, D. O., *Effects of Forest Insect Spraying on Trout and Aquatic Insects in Some Montana Streams*. Final Report, Montana Fish and Game Dept, 1958.

PAGES 128–9 Graham, R. J., 'Effects of Forest Insect Spraying on Trout and Aquatic Insects in Some Montana Streams', in *Biological Problems in Water Pollution*. Transactions, 1959 seminar. U.S. Public Health Service Technical Report W60–3 (1960).

PAGE 129 Crouter, R. A. and Vernon, E. H., 'Effects of Black-headed Budworm Control on Salmon and Trout in British Columbia', *Canadian Fish Culturist*, Issue 24 (1959), pp. 23–40.

PAGES 129–30 *Pollution-Caused Fish Kills in 1960*. U.S. Public Health Service Publ. No. 847 (1961), pp. 1–20.

PAGE 130 Whiteside, J. M., 'Spruce Budworm Control in Oregon and Washington, 1949–1956', *Proc.*, 10th Internatl Congress of Entomologists (1956), Vol. 4 (1958), pp. 291–302.

PAGES 130–1 'U.S. Anglers – Three Billion Dollars', *Sport Fishing Inst. Bull.*, No. 119 (October 1961).

PAGE 131 Powers, Edward (Bur. of Commercial Fisheries), To author.

PAGE 131   Rudd, Robert L. and Genelly, Richard E., *Pesticides: Their Use and Toxicity in Relation to Wildlife*. Calif. Dept of Fish and Game, Game Bulletin No. 7 (1956), p. 88.

PAGE 131   Biglane, K. E., To author, 8 May 1961.

PAGE 131   Release No. 58-38, Pennsylvania Fish Commission, 8 December 1958.

PAGE 131   Rudd, Robert L. and Genelly, Richard E., *Pesticides*, p. 60.

PAGE 131   Henderson, C., *et al.*, 'The Relative Toxicity of Ten Chlorinated Hydrocarbon Insecticides to Four Species of Fish', paper presented at 88th Annual Meeting Am. Fisheries Soc. (1958).

PAGE 131   'The Fire Ant Eradication Program and How It Affects Wildlife', subject of *Proc. Symposium*, 12th Annual Conf. South-eastern Assn Game and Fish Commissioners, Louisville, Ky (1958). Pub. by the Assn, Columbia, S.C., 1958.

PAGE 131   'Effects of the Fire Ant Eradication Program on Wildlife', report, U.S. Fish and Wildlife Service, 25 May 1958. Mimeo.

PAGE 131   *Pesticide-Wildlife Review*, 1959. Bur. Sport Fisheries and Wildlife Circular 84 (1960), U.S. Fish and Wildlife Service, pp. 1-36.

PAGE 131   Baker, Maurice F., 'Observations of Effects of an Application of Heptachlor or Dieldrin on Wildlife', in *Proc. Symposium*, pp. 18-20.

PAGES 131-2   Glasgow, L. L., 'Studies on the Effect of the Imported Fire Ant Control Program on Wildlife in Louisiana', in *Proc. Symposium*, pp. 24-9.

PAGE 132   *Pesticide-Wildlife Review*, 1959.

PAGE 132   *Progress in Sport Fishery Research*, 1960. Bur. Sport Fisheries and Wildlife Circular 101 (1960), U.S. Fish and Wildlife Service.

PAGE 132   'Resolution Opposing Fire-Ant Program Passed by American Society of Ichthyologists and Herpetologists', *Copeia* (1959), No. 1, p. 89.

PAGE 132   Young, L. A. and Nicholson, H. P., 'Stream Pollution Resulting from the Use of Organic Insecticides', *Progressive Fish Culturist*, Vol. 13 (1951), No. 4, pp. 193-8.

PAGE 134 Rudd, Robert L. and Genelly, Richard E., *Pesticides*.

PAGE 134 Lawrence, J. M., 'Toxicity of Some New Insecticides to Several Species of Pondfish', *Progressive Fish Culturist*, Vol. 12 (1950), No. 4, pp. 141–6.

PAGE 134 Pielow, D. P., 'Lethal Effects of DDT on Young Fish', *Nature*, Vol. 158 (1946), No. 4011, p. 378.

PAGE 135 Herald, E. S., 'Notes on the Effect of Aircraft-Distributed DDT-Oil Spray upon Certain Philippine Fishes', *Jour. Wildlife Management*, Vol. 13 (1949), No. 3, p. 316.

PAGES 135–6 'Report of Investigation of the Colorado River Fish Kill, January, 1961'. Texas Game and Fish Commission, 1961. Mimeo.

PAGE 137 Harrington, R. W., Jr, and Bidlingmayer, W. L., 'Effects of Dieldrin on Fishes and Invertebrates of a Salt Marsh', *Jour. Wildlife Management*, Vol. 22 (1958), No. 1, pp. 76–82.

PAGES 137–8 Mills, Herbert R., 'Death in the Florida Marshes', *Audubon Mag.*, September–October 1952.

PAGE 138 Springer, Paul F. and Webster, John R., *Effects of DDT on Saltmarsh Wildlife: 1949*. U.S. Fish and Wildlife Service, Special Scientific Report, Wildlife No. 10 (1949).

PAGE 139 John C. Pearson, To author.

PAGES 140–1 Butler, Philip A., 'Effects of Pesticides on Commercial Fisheries', *Proc.*, 13th Annual Session (November 1960), Gulf and Caribbean Fisheries Inst., pp. 168–71.

CHAPTER 10: INDISCRIMINATELY FROM THE SKIES

PAGE 142 Perry, C. C., *Gypsy Moth Appraisal Program and Proposed Plan to Prevent Spread of the Moths*. U.S. Dept of Agric. Technical Bulletin No. 1124 (October 1955).

PAGE 143 Corliss, John M., 'The Gypsy Moth', *Yearbook of Agric.*, U.S. Dept of Agric., 1952, pp. 694–8.

PAGE 143 Worrell, Albert C., 'Pests, Pesticides and People', offprint from *Am. Forests Mag.*, July 1960.

PAGE 143 Clausen, C. P., 'Parasites and Predators', *Yearbook of Agric.*, U.S. Dept of Agric., 1952, pp. 380–8.

PAGE 143   Perry, C. C., *Gypsy Moth Appraisal Program.*

PAGE 144   Worrell, Albert C., 'Pests, Pesticides, and People'.

PAGE 144   'USDA Launches Large-Scale Effort to Wipe Out Gypsy Moth', press release, U.S. Dept of Agric., 20 March 1957.

PAGE 144   Worrell, Albert C., 'Pests, Pesticides, and People'.

PAGE 144   *Robert Cushman Murphy et al.* v. *Ezra Taft Benson et al.* U.S. District Court, Eastern District of New York, October 1959, Civ. No. 17610.

PAGE 145   *Murphy et al.* v. *Benson et al.* Petition for a Writ of Certiorari to the U.S. Court of Appeals for the Second Circuit, October 1959.

PAGE 145   Waller, W. K., 'Poison on the Land', *Audubon Mag.*, March–April 1958, pp. 68–71.

PAGE 145   *Murphy et al.* v. *Benson et al.* U.S. Supreme Court Reports, Memorandum Cases, No. 662, 28 March 1960.

PAGE 145   Waller, W. K., 'Poison on the Land'.

PAGE 146   *Am. Bee Jour.*, June 1958, p. 224.

PAGE 147   *Murphy et al.* v. *Benson et al.* U.S. Court of Appeals, Second Circuit. Brief for Defendant-Appellee Butler, No. 25,448, March 1959.

PAGE 147   Brown, William L., Jr, 'Mass Insect Control Programs: Four Case Histories', *Psyche*, Vol. 68 (1961), Nos. 2–3, pp. 75–111.

PAGES 147–8   Arant, F. S., *et al.*, 'Facts about the Imported Fire Ant', *Highlights of Agric. Research*, Vol. 5 (1958), No. 4.

PAGE 148   Brown, William L., Jr, 'Mass Insect Control Programs'.

PAGE 148   'Pesticides: Hedgehopping into Trouble?' *Chemical Week*, 8 February 1958, p. 97.

PAGE 149   Arant, F. S., *et al.*, 'Facts about the Imported Fire Ant'.

PAGE 149   Byrd, I. B., 'What Are the Side Effects of the Imported Fire Ant Control Program?' in *Biological Problems in Water Pollution.* Transactions, 1959 seminar, U.S. Public Health Service Technical Report W60-3 (1960), pp. 46–50.

PAGE 149   Hays, S. B. and Hays, K. L., 'Food Habits of *Solenopsis saevissima richteri* Forel', *Jour. Econ. Entomol.*, Vol. 52 (1959), No. 3, pp. 455–7.

PAGE 149 Caro, M. R., *et al.*, 'Skin Responses to the Sting of the Imported Fire Ant', *A.M.A. Archives Dermat.*, Vol. 75 (1957), pp. 475–88.

PAGE 149 Byrd, I. B., 'Side Effects of Fire Ant Program'.

PAGE 150 Baker, Maurice F., in *Virginia Wildlife*, November 1958.

PAGES 150–1 'The Fire Ant Eradication Program and How It Affects Wildlife', subject of *Proc. Symposium*, 12th Annual Conf. South-eastern Assn Game and Fish Commissioners, Louisville, Ky (1958). Published by the Assn, Columbia, S.C., 1958.

PAGE 151 Brown, William L., Jr, 'Mass Insect Control Programs'.

PAGE 151 *Pesticide-Wildlife Review*, 1959. Bur. Sport Fisheries and Wildlife Circular 84 (1960), U.S. Fish and Wildlife Service, pp. 1–36.

PAGES 151–2 Wright, Bruce S., 'Woodcock Reproduction in DDT-Sprayed Areas of New Brunswick', *Jour. Wildlife Management*, Vol. 24 (1960), No. 4, pp. 419–20.

PAGE 152 Clawson, Sterling G., 'Fire Ant Eradication – and Quail', *Alabama Conservation*, Vol. 30 (1959). No. 4, p. 14.

PAGE 152 Rosene, Walter, 'Whistling-Cock Counts of Bob-white Quail on Areas Treated with Insecticide and on Un-treated Areas, Decatur County, Georgia', in *Proc. Symposium*, pp. 14–18.

PAGE 152 *Pesticide-Wildlife Review*, 1959.

PAGES 152–3 Cottam, Clarence, 'The Uncontrolled Use of Pesticides in the South-east', address to South-eastern Assn Fish, Game and Conservation Commissioners, October 1959.

PAGES 153–4 Poitevint, Otis L., Address to Georgia Sportsmen's Fed., October 1959.

PAGE 154 Ely, R. E., *et al.*, 'Excretion of Heptachlor Epoxide in the Milk of Dairy Cows Fed Heptachlor-Sprayed Forage and Technical Heptachlor', *Jour. Dairy Sci.*, Vol. 38 (1955), No. 6, pp. 669–72.

PAGE 154 Gannon, N., *et al.*, 'Storage of Dieldrin in Tissues and Its Excretion in Milk of Dairy Cows Fed Dieldrin in Their Diets', *Jour. Agric. and Food Chem.*, Vol. 7 (1959), No. 12, pp. 824–32.

PAGE 154　*Insecticide Recommendations of the Entomology Research Division for the Control of Insects Attacking Crops and Livestock for 1961.* U.S. Dept of Agric. Handbook No. 120 (1961).

PAGE 154　Peckinpaugh, H. S. (Alabama Dept of Agric. and Indus.), To author, 24 March 1959.

PAGE 154　Hartman, H. L. (Louisiana State Board of Health), To author, 23 March 1959.

PAGE 154　Lakey, J. F. (Texas Dept of Health), To author, 23 March 1959.

PAGE 155　Davidow, B. and Radomski, J. L., 'Metabolite of Heptachlor, Its Analysis, Storage, and Toxicity', *Federation Proc.*, Vol. 11 (1952), No. 1, p. 336.

PAGE 155　Food and Drug Administration, U.S. Dept of Health, Education, and Welfare, in *Federal Register*, 27 October 1959.

PAGE 155　Burgess, E. D. (U.S. Dept of Agric.), To author, 23 June 1961.

PAGE 156　'Fire Ant Control is Parley Topic', *Beaumont [Texas] Journal*, 24 September 1959.

PAGE 156　'Coordination of Pesticides Programs', *Hearings*, 86th Congress, H.R. 11502, Com. on Merchant Marine and Fisheries, May 1960, p. 45.

PAGE 156　Newsom, L. D. (Head, Entomol. Research, Louisiana State Univ.), To author, 23 March 1962.

PAGE 157　Green, H. B. and Hutchins, R. E., *Economical Method for Control of Imported Fire Ant in Pastures and Meadows.* Miss. State Univ. Agric. Exper. Station Information Sheet 586 (May 1958).

CHAPTER II: BEYOND THE DREAMS OF THE BORGIAS

PAGE 159　'Chemicals in Food Products', *Hearings*, 81st Congress, H.R. 323, Com. to Investigate Use of Chemicals in Food Products, Pt 1 (1950), pp. 388–90.

PAGE 160　*Clothes Moths and Carpet Beetles.* U.S. Dept of Agric., Home and Garden Bulletin No. 24 (1961).

PAGE 160 Mulrennan, J. A., To author, 15 March 1960.

PAGE 160 *New York Times*, 22 May 1960.

PAGE 161 Petty, Charles S., 'Organic Phosphate Insecticide Poisoning. Residual Effects in Two Cases', *Am. Jour. Med.*, Vol. 24 (1958), pp. 467–70.

PAGE 161 Miller, A. C., *et al.*, 'Do People Read Labels on Household Insecticides?' *Soap and Chem. Specialties*, Vol. 34 (1958), No. 7, pp. 61–3.

PAGE 162 Hayes, Wayland J., Jr, *et al.*, 'Storage of DDT and DDE in People with Different Degrees of Exposure to DDT', *A.M.A. Archives Indus. Health*, Vol. 18 (November 1958), pp. 398–406.

PAGE 162 Walker, Kenneth C., *et al.*, 'Pesticide Residues in Foods. Dichlorodiphenyltrichloroethane and Dichlorodiphenyldichloroethylene Content of Prepared Meals', *Jour. Agric. and Food Chem.*, Vol. 2 (1954), No. 20, pp. 1034–7.

PAGE 162 Hayes, Wayland J., Jr, *et al.*, 'The Effect of Known Repeated Oral Doses of Chlorophenothane (DDT) in Man', *Jour. Am. Med. Assn.*, Vol. 162 (1956), No. 9, pp. 890–97.

PAGE 162 Milstead, K. L., 'Highlights in Various Areas of Enforcement', address to 64th Annual Conf. Assn of Food and Drug Officials of U.S., Dallas (June 1960).

PAGE 163 Durham, William, *et al.*, 'Insecticide Content of Diet and Body Fat of Alaskan Natives', *Science*, Vol. 134 (1961), No. 3493, pp. 1880–1.

PAGE 164 'Pesticides – 1959', *Jour. Agric. and Food Chem.*, Vol. 7 (1959), No. 10, pp. 674–88.

PAGE 164 *Annual Reports*, Food and Drug Administration, U.S. Dept of Health, Education, and Welfare. For 1957, pp. 196, 197; 1956, p. 203.

PAGE 164 Markarian, Haig, *et al.*, 'Insecticide Residues in Foods Subjected to Fogging under Simulated Warehouse Conditions', *Abstracts*, 135th Meeting Am. Chem. Soc. (April 1959).

CHAPTER 12: THE HUMAN PRICE

PAGE 168 Price, David E., 'Is Man Becoming Obsolete?' *Public Health Reports*, Vol. 74 (1959), No. 8, pp. 693–9.

PAGE 169 'Report on Environment Health Problems', *Hearings*, 86th Congress, Subcom. of Com. on Appropriations, March 1960, p. 34.

PAGE 169 Dubos, René, *Mirage of Health*, New York, Harper, 1959. World Perspectives Series, P. 171; London, Allen & Unwin, 1960.

PAGE 169 *Medical Research: A Midcentury Survey*. Vol. 2, *Unsolved Clinical Problems in Biological Perspective*, Boston, Little, Brown, 1955, p. 4.

PAGE 170 'Chemicals in Food Products', *Hearings*, 81st Congress, H.R. 323, Com. to Investigate Use of Chemicals in Food Products, 1950, p. 5. Testimony of A. J. Carlson.

PAGE 170 Paul, A. H., 'Dieldrin Poisoning – A Case Report', *New Zealand Med. Jour.*, Vol. 58 (1959), p. 393.

PAGE 170 'Insecticide Storage in Adipose Tissue', editorial, *Jour. Am. Med. Assn.*, Vol. 145 (10 March 1951), pp. 735–6.

PAGE 171 Mitchell, Philip H., *A Textbook of General Physiology*, New York, McGraw-Hill, 1956, 5th ed.

PAGE 171 Miller, B. F. and Goode, R., *Man and His Body: The Wonders of the Human Mechanism*, New York, Simon and Schuster, 1960; London, Gollancz, 1961.

PAGE 171 Dubois, Kenneth P., 'Potentiation of the Toxicity of Insecticidal Organic Phosphates', *A.M.A. Archives Indus. Health*, Vol. 18 (December 1958), pp. 488–96.

PAGE 172 Gleason, Marion, *et al.*, *Clinical Toxicology of Commercial Products*, Baltimore, Williams and Wilkins, 1957.

PAGES 172–3 Case, R. A. M., 'Toxic Effects of DDT in Man', *Brit. Med. Jour.*, Vol. 2 (15 December 1945), pp. 842–5.

PAGE 173 Wigglesworth, V. D., 'A Case of DDT Poisoning in Man', *Brit. Med. Jour.*, Vol. 1 (14 April 1945), p. 517.

PAGE 173 Hayes, Wayland J., Jr, *et al.*, 'The Effect of Known Repeated Oral Doses of Chlorophenothane (DDT) in Man', *Jour. Am. Med. Assn.*, Vol. 162 (27 October 1956), pp. 890–7.

PAGE 173 Hargraves, Malcolm M., 'Chemical Pesticides and

Conservation Problems', address to 23rd Annual Conv. Natl Wildlife Fed. (27 February 1959). Mimeo.

PAGE 174 Hargraves, Malcolm M. and Hanlon, D. G., 'Leukemia and Lymphoma – Environmental Diseases?' paper presented at Internatl Congress of Hematology, Japan, September 1960. Mimeo.

PAGE 174 'Chemicals in Food Products', *Hearings*, 81st Congress, H.R. 323, Com. to Investigate Use of Chemicals in Food Products, 1950. Testimony of Dr Morton S. Biskind.

PAGE 174 Thompson, R. H. S., 'Cholinesterases and Anticholinesterases', *Lectures on the Scientific Basis of Medicine*, Vol. II (1952–3), Univ. of London. London, Athlone Press, 1954.

PAGE 175 Laug, E. P. and Keenz, F. M., 'Effect of Carbon Tetrachloride on Toxicity and Storage of Methoxychlor in Rats', *Federation Proc.*, Vol. 10 (March 1951), p. 318.

PAGE 175 Hayes, Wayland J., Jr, 'The Toxicity of Dieldrin to Man', *Bull. World Health Organ.*, Vol. 20 (1959), pp. 891–912.

PAGE 175 'Abuse of Insecticide Fumigating Devices', *Jour. Am. Med. Assn.*, Vol. 156 (9 October 1954), pp. 607–8.

PAGE 176 'Chemicals in Food Products', Testimony of Dr Paul B. Dunbar, pp. 28–9.

PAGE 176 Smith, M. I. and Elrove, E., 'Pharmacological and Chemical Studies of the Cause of So-Called Ginger Paralysis', *Public Health Reports*, Vol. 45 (1930), pp. 1703–16.

PAGE 176 Durham, W. F., *et al.*, 'Paralytic and Related Effects of Certain Organic Phosphorus Compounds', *A.M.A. Archives Indus. Health*, Vol. 13 (1956), pp. 326–30.

PAGE 176 Bidstrup, P. L., *et al.*, 'Anticholinesterases (Paralysis in Man Following Poisoning by Cholinesterase Inhibitors)', *Chem. and Indus.*, Vol. 24 (1954), pp. 674–6.

PAGE 177 Gershon, S. and Shaw, F. H., 'Psychiatric Sequelae of Chronic Exposure to Organophosphorus Insecticides', *Lancet*, Vol. 7191 (24 June 1961), pp. 1371–4.

CHAPTER 13: THROUGH A NARROW WINDOW

PAGE 178 Wald, George, 'Life and Light', *Sci. American*, October 1959, pp. 40–2.

PAGE 178 Rabinowitch, E. I., Quoted in *Medical Research: A*

Midcentury Survey. Vol. 2, *Unsolved Clinical Problems in Biological Perspective*, Boston, Little, Brown, 1955, p. 25.

PAGE 179   Ernster, L. and Lindberg, O., 'Animal Mitochondria', *Annual Rev. Physiol.*, Vol. 20 (1958), pp. 13–42.

PAGE 180   Siekevitz, Philip, 'Powerhouse of the Cell', *Sci. American*, Vol. 197 (1957), No. 1, pp. 131–40.

PAGE 180   Green, David E., 'Biological Oxidation', *Sci. American*, Vol. 199 (1958), No. 1, pp. 56–62.

PAGE 180   Lehninger, Albert L., 'Energy Transformation in the Cell', *Sci. American*, Vol. 202 (1960), No. 5, pp. 102–14.

PAGE 180   Lehninger, Albert L., *Oxidative Phosphorylation*. Harvey Lectures (1953–54), Ser. XLIX, Harvard University. Cambridge, Harvard Univ. Press, 1955, pp. 176–215.

PAGE 181   Siekevitz, Philip, 'Powerhouse of the Cell'.

PAGE 181   Simon, E. W., 'Mechanisms of Dinitrophenol Toxicity', *Biol. Rev.*, Vol. 28 (1953), pp. 453–79.

PAGE 181   Yost, Henry T. and Robson, H. H., 'Studies on the Effects of Irradiation of Cellular Particulates. III. The Effect of Combined Radiation Treatments on Phosphorylation', *Biol. Bull.*, Vol. 116 (1959), No. 3, pp. 498–506.

PAGE 181   Loomis, W. F. and Lipmann, F., 'Reversible Inhibition of the Coupling between Phosphorylation and Oxidation', *Jour. Biol. Chem.*, Vol. 173 (1948), pp. 807–8.

PAGE 182   Brody, T. M., 'Effect of Certain Plant Growth Substances on Oxidative Phosphorylation in Rat Liver Mitochondria', *Proc. Soc. Exper. Biol. and Med.*, Vol. 80 (1952), pp. 533–6.

PAGE 182   Sacklin, J. A., *et al.*, 'Effect of DDT on Enzymatic Oxidation and Phosphorylation', *Science*, Vol. 122 (1955), pp. 377–8.

PAGE 182   Danziger, L., 'Anoxia and Compounds Causing Mental Disorders in Man', *Diseases Nervous System*, Vol. 6 (1945), No. 12, pp. 365–70.

PAGE 182   Goldblatt, Harry and Cameron G., 'Induced Malignancy in Cells from Rat Myocardium Subjected to Intermittent Anaerobiosis During Long Propagation in Vitro', *Jour. Exper. Med.*, Vol. 97 (1953), No. 4, pp. 525–52.

PAGE 182   Warburg, Otto, 'On the Origin of Cancer Cells', *Science*, Vol. 123 (1956), No. 3191, pp. 309–14.

PAGES 182–3 'Congenital Malformations Subject of Study', *Registrar*, U.S. Public Health Service, Vol. 24, No. 12 (December 1959), p. 1.

PAGE 183 Brachet, J., *Biochemical Cytology*, New York, Academic Press, 1957, p. 516.

PAGE 184 Genelly, Richard E. and Rudd, Robert L., 'Effects of DDT, Toxaphene, and Dieldrin on Pheasant Reproduction', *Auk*, Vol. 73 (October 1956), pp. 529–39.

PAGE 184 Wallace, George J., To author, 2 June 1960.

PAGE 184 Cottam, Clarence, 'Some Effects of Sprays on Crops and Livestock', address to Soil Conservation Soc. of Am., August 1961. Mimeo.

PAGE 184 Bryson, M. J., *et al.*, 'DDT in Eggs and Tissues of Chickens Fed Varying Levels of DDT', *Advances in Chem.*, Ser. No. 1, 1950.

PAGE 184 Pillmore, R. E., 'Insecticide Residues in Big Game Animals', U.S. Fish and Wildlife Service, pp. 1–10. Denver, 1961. Mimeo.

PAGE 185 Genelly, Richard E. and Rudd, Robert L., 'Chronic Toxicity of DDT, Toxaphene, and Dieldrin to Ring-necked Pheasants', *Calif. Fish and Game*, Vol. 42 (1956), No. 1, pp. 5–14.

PAGE 185 Emmel, L. and Krupe, M., 'The Mode of Action of DDT in Warm-blooded Animals', *Zeits. fur Naturforschung*, Vol. 1 (1946), pp. 691–5.

PAGE 185 Wallace, George J., To author.

PAGE 185 Hodge, C. H., *et al.*, 'Short-Term Oral Toxicity Tests of Methoxychlor in Rats and Dogs', *Jour. Pharmacol. and Exper. Therapeut.*, Vol. 99 (1950), p. 140.

PAGE 185 Burlington, H. and Lindeman, V. F., 'Effect of DDT on Testes and Secondary Sex Characters of White Leghorn Cockerels', *Proc. Soc. Exper. Biol. and Med.*, Vol. 74 (1950), pp. 48–51.

PAGE 185 Lardy, H. A. and Phillips, P. H., 'The Effect of Thyroxine and Dinitrophenol on Sperm Metabolism', *Jour. Biol. Chem.*, Vol. 149 (1943), p. 177.

PAGE 185 'Occupational Oligospermia', letter to Editor, *Jour. Am. Med. Assn.*, Vol. 140, No. 1249 (13 August 1949).

PAGE 18 Burnet, F. Macfarlane, 'Leukemia As a Problem in

Preventive Medicine', *New Eng. Jour. Med.*, Vol. 259 (1958), No. 9, pp. 423–31.

PAGE 185 Alexander, Peter, 'Radiation-Imitating Chemicals', *Sci. American*, Vol. 202 (1960), No. 1, pp. 99–108.

PAGE 187 Simpson, George G., Pittendrigh, C. S. and Tiffany, L. H., *Life: An Introduction to Biology*, New York, Harcourt, Brace, 1957; London, Routledge & Kegan Paul, 1958.

PAGE 188 Burnet, F., 'Leukemia As a Problem in Preventive Medicine'.

PAGE 188 Bearn, A. G. and German III, J. L., 'Chromosomes and Disease', *Sci. American*, Vol. 205 (1961), No. 5, pp. 66–76.

PAGE 188 'The Nature of Radioactive Fall-out and Its Effects on Man', *Hearings*, 85th Congress, Joint Com. on Atomic Energy, Pt 2 (June 1957), p. 1062. Testimony of Dr Hermann J. Muller.

PAGE 188 Alexander, Peter, 'Radiation-Imitating Chemicals'.

PAGE 188 Muller, Hermann J., 'Radiation and Human Mutation', *Sci. American*, Vol. 193 (1955), No. 11, pp. 58–68.

PAGE 189 Conen, P. E. and Lansky, G. S., 'Chromosome Damage during Nitrogen Mustard Therapy', *Brit. Med. Jour.*, Vol. 2 (21 October 1961), pp. 1055–7.

PAGE 189 Blasquez, J. and Maier, J., 'Ginandromorfismo en *Culex fatigans* sometidos por generaciones sucesivas a exposiciones de DDT', *Revista de Sanidad y Assistencia Social* (Caracas), Vol. 16 (1951), pp. 607–12.

PAGE 189 Levan, A. and Tjio, J. H., 'Induction of Chromosome Fragmentation by Phenols', *Hereditas*, Vol. 34 (1948), pp. 453–84.

PAGE 189 Loveless, A. and Revell, S., 'New Evidence on the Mode of Action of "Mitotic Poisons"', *Nature*, Vol. 164 (1949), pp. 938–44.

PAGE 189 Hadorn, E., *et al.* Quoted by Charlotte Auerbach in 'Chemical Mutagenesis', *Biol. Rev.*, Vol. 24 (1949), pp. 355–91.

PAGE 189 Wilson, S. M., *et al.*, 'Cytological and Genetical Effects of the Defoliant Endothal', *Jour. of Heredity*, Vol. 47 (1956), No. 4, pp. 151–5.

PAGE 189 Vogt, quoted by W. J. Burdette in 'The Significance of Mutation in Relation to the Origin of Tumors: A Review', *Cancer Research*, Vol. 15 (1955), No. 4, pp. 201–26.

PAGE 189 Swanson, Carl, *Cytology and Cytogenetics*, Englewood Cliffs, N.J., Prentice-Hall, 1957.

PAGE 189 Kostoff, D., 'Induction of Cytogenic Changes and Atypical Growth by Hexachlorcyclohexane', *Science*, Vol. 109 (6 May 1949), pp. 467–8.

PAGE 189 Sass, John E., 'Response of Meristems of Seedlings to Benzene Hexachloride Used As a Seed Protectant', *Science*, Vol. 114 (2 November 1951), p. 466.

PAGE 190 Shenefelt, R. D., 'What's Behind Insect Control?' in *What's New in Farm Science*. Univ. of Wisc. Agric. Exper. Station Bulletin 512 (January 1955).

PAGE 190 Croker, Barbara H., 'Effects of 2,4-D and 2,4,4-T on Mitosis in *Allium cepa*', *Bot. Gazette*, Vol. 114 (1953), pp. 274–83.

PAGE 190 Mühling, G. N., *et al.*, 'Cytological Effects of Herbicidal Substituted Phenols', *Weeds*, Vol. 8 (1960), No. 2, pp. 173–81.

PAGE 190 Davis, David E., To author, 24 November 1961.

PAGE 190 Jacobs, Patricia A., *et al.*, 'The Somatic Chromosomes in Mongolism', *Lancet*, No. 7075 (4 April 1959), p. 710.

PAGE 190 Ford, C. E. and Jacobs, P. A., 'Human Somatic Chromosomes', *Nature*, 7 June 1958, pp. 1565–8.

PAGE 191 'Chromosome Abnormality in Chronic Myeloid Leukaemia', editorial, *Brit. Med. Jour.*, Vol. 1 (4 February 1961), p. 347.

PAGE 191 Bearn, A. G. and German III, J. L., 'Chromosomes and Disease'.

PAGE 191 Patau, K., *et al.*, 'Partial-Trisomy Syndromes. I. Sturge-Weber's Disease', *Am. Jour. Human Genetics*, Vol. 13 (1961), No. 3, pp. 287–98.

PAGE 192 Patau, K., *et al.*, 'Partial-Trisomy Syndromes. II. An Insertion As Cause of the OFD Syndrome in Mother and Daughter', *Chromosoma* (Berlin), Vol. 21 (1961), pp. 573–84.

PAGE 192 Therman, E., *et al.*, 'The D Trisomy Syndrome and XO Gonadal Dysgenesis in Two Sisters', *Am. Jour. Human Genetics*, Vol. 13 (1961), No. 2, pp. 193–204.

CHAPTER 14: ONE IN EVERY FOUR

PAGE 193 Hueper, W. C., 'Newer Developments in Occupational and Environmental Cancer', *A.M.A. Archives Inter. Med.*, Vol. 100 (September 1957), pp. 487–503.

PAGE 194 Hueper, W. C., *Occupational Tumors and Allied Diseases*, Springfield, Ill., Thomas, 1942.

PAGE 195 Hueper, W. C., 'Environmental Cancer Hazards: A Problem of Community Health', *Southern Med. Jour.*, Vol. 50 (1957), No. 7, pp. 923–33.

PAGE 195 'Estimated Numbers of Deaths and Death Rates for Selected Causes: United States', Annual Summary for 1959, Pt 2, *Monthly Vital Statistics Report*, Vol. 7, No. 13 (22 July 1959), p. 14. Natl Office of Vital Statistics, Public Health Service.

PAGE 195 *1962 Cancer Facts and Figures.* American Cancer Society.

PAGE 195 *Vital Statistics of the United States, 1959.* Natl Office of Vital Statistics, Public Health Service, Vol. I, Sec. 6, Mortality Statistics. Table 6-K.

PAGE 195 Hueper, W. C., *Environmental and Occupational Cancer*, Public Health Reports, Supplement 209 (1948).

PAGE 195 'Food Additives', *Hearings*, 85th Congress, Subcom. of Com. on Interstate and Foreign Commerce, 19 July 1957. Testimony of Dr Francis E. Ray, p. 200.

PAGE 197 Hueper, W. C., *Occupational Tumors and Allied Diseases*.

PAGE 197 Hueper, W. C., 'Potential Role of Non-Nutritive Food Additives and Contaminants as Environmental Carcinogens', *A.M.A. Archives Path.*, Vol. 62 (September 1956), pp. 218–49.

PAGE 198 'Tolerances for Residues or Aramite', *Federal Register*, 30 September 1955. Food and Drug Administration, U.S. Dept of Health, Education, and Welfare.

PAGE 198 'Notice of Proposal to Establish Zero Tolerances for Aramite', *Federal Register*, 26 April 1958. Food and Drug Administration.

PAGE 198 'Aramite – Revocation of Tolerances; Establishment of Zero Tolerances', *Federal Register*, 24 December 1958. Food and Drug Administration.

PAGE 198   Van Oettingen, W. F., *The Halogenated Aliphatic, Olefinic, Cyclic, Aromatic, and Aliphatic-Aromatic Hydrocarbons: Including the Halogenated Insecticides, Their Toxicity and Potential Dangers*. U.S. Dept of Health, Education, and Welfare. Public Health Service Publ. No. 414 (1955).

PAGE 198   Hueper, W. C. and Payne, W. W., 'Observations on the Occurrence of Hepatomas in Rainbow Trout', *Jour. Natl. Cancer Inst.*, Vol. 27 (1961), pp. 1123–43.

PAGE 198   VanEsch, G. J., *et al.*, 'The Production of Skin Tumours in Mice by Oral Treatment with Urethane-Isopropyl-N-Phenyl Carbamate or Isopropyl-N-Chlorophenyl Carbamate in Combination with Skin Painting with Croton Oil and Tween 60', *Brit. Jour. Cancer*, Vol. 12 (1958), pp. 355–62.

PAGES 198–9   'Scientific Background for Food and Drug Administration Action against Aminotriazole in Cranberries', Food and Drug Administration, U.S. Dept of Health, Education, and Welfare, 17 November 1959. Mimeo.

PAGE 199   Rutstein, David. Letter to *New York Times*, 16 November 1959.

PAGE 199   Hueper, W. C., 'Causal and Preventive Aspects of Environmental Cancer', *Minnesota Med.*, Vol. 39 (January 1956), pp. 5–11, 22.

PAGE 200   'Estimated Numbers of Deaths and Death Rates for Selected Causes: United States', Annual Summary for 1960, Pt 2, *Monthly Vital Statistics Report*, Vol. 9, No. 13 (28 July 1961), Table 3.

PAGE 200   *Robert Cushman Murphy et al.* v. *Ezra Taft Benson et al.* U.S. District Court, Eastern District of New York, October 1959, Civ. No. 17610. Testimony of Dr Malcolm M. Hargraves.

PAGES 200–1   Hargraves, Malcolm M., 'Chemical Pesticides and Conservation Problems', address to 23rd Annual Conv. Natl Wildlife Fed. (27 February 1959). Mimeo.

PAGE 202   Hargraves, Malcolm M. and Hanlon, D. G., 'Leukemia and Lymphoma – Environmental Diseases?' paper presented at Internatl Congress of Hematology, Japan, September 1960. Mimeo.

PAGE 202 Wright, C., *et al.*, 'Agranulocytosis Occurring after Exposure to a DDT Pyrethrum Aerosol Bomb', *Am. Jour. Med.*, Vol. 1 (1946), pp. 562–7.

PAGE 202 Jedlicka, V., 'Paramyeloblastic Leukemia Appearing Simultaneously in Two Blood Cousins after Simultaneous Contact with Gammexane (Hexachlorcyclohexane)', *Acta Med. Scand.*, Vol. 161 (1958), pp. 447–51.

PAGE 202 Friberg, L. and Martensson, J., 'Case of Panmyelopthisis after Exposure to Chlorophenothane and Benzene Hexachloride', (A.M.A.) *Archives Indus. Hygiene and Occupat. Med.*, Vol. 8 (1953), No. 2, pp. 166–9.

PAGES 203–4 Warburg, Otto, 'On the Origin of Cancer Cells', *Science*, Vol. 123, No. 3191 (24 February 1956), pp. 309–14.

PAGE 205 Sloan-Kettering Inst. for Cancer Research, *Biennial Report*, 1 July 1957–30 June 1959, p. 72.

PAGES 205–6 Levan, Albert and Biesele, John J., 'Role of Chromosomes in Cancerogenesis ,as Studied in Serial Tissue Culture of Mammalian Cells', *Annals New York Acad. Sci.*, Vol. 71 (1958), No. 6, pp. 1022–53.

PAGE 206 Hunter, F. T., 'Chronic Exposure to Benzene (Benzol). II. The Clinical Effects', *Jour. Indus. Hygiene and Toxicol.*, Vol. 21 (1939), pp. 331–54.

PAGE 206 Mallory, T. B., *et al.*, 'Chronic Exposure to Benzene (Benzol). III. The Pathologic Results', *Jour. Indus. Hygiene and Toxicol.*, Vol. 21 (1939), pp. 355–93.

PAGE 206 Hueper, W. C., *Environmental and Occupational Cancer*, pp. 1–69.

PAGE 206 Hueper, W. C., 'Recent Developments in Environmental Cancer', *A.M.A. Archives Path.*, Vol. 58 (1954), pp. 475–523.

PAGES 206–7 Burnet, F. Macfarlane, 'Leukemia As a Problem in Preventive Medicine', *New Eng. Jour. Med.*, Vol. 259 (1958), No. 9, pp. 423–31.

PAGE 207 Klein, Michael, 'The Transplacental Effect of Urethan on Lung Tumorigenesis in Mice', *Jour. Natl. Cancer Inst.*, Vol. 12 (1952), pp. 1003–10.

PAGES 207–9 Biskind, M. S. and Biskind, G. R., 'Diminution in Ability of the Liver to Inactivate Estrone in Vitamin B

Complex Deficiency', *Science*, Vol. 94, No. 2446 (November 1941), p. 462.

PAGES 207–9 Biskind, G. R. and Biskind, M. S., 'The Nutritional Aspects of Certain Endocrine Disturbances', *Am. Jour. Clin. Path.*, Vol. 16 (1946), No. 12, pp. 737–45.

PAGES 207–9 Biskind, M. S. and Biskind, G. R., 'Effect of Vitamin B Complex Deficiency on Inactivation of Estrone in the Liver' *Endocrinology*, Vol. 31 (1942), No. 1, pp. 109–14.

PAGES 207–9 Biskind, M. S. and Shelesnyak, M. C., 'Effect of Vitamin B Complex Deficiency on Inactivation of Ovarian Estrogen in the Liver', *Endocrinology*, Vol. 30 (1942), No. 5, pp. 819–20.

PAGES 207–9 Biskind, M. S. and Biskind, G. R., 'Inactivation of Testosterone Propionate in the Liver During Vitamin B Complex Deficiency. Alteration of the Estrogen-Androgen Equilibrium', *Endocrinology*, Vol. 32 (1943), No. 1, pp. 97–102.

PAGE 208 Greene, H. S. N., 'Uterine Adenomata in the Rabbit. III. Susceptibility As a Function of Constitutional Factors', *Jour. Exper. Med.*, Vol. 73 (1941), No. 2, pp. 273–92.

PAGE 208 Horning, E. S. and Whittick, J. W., 'The Histogenesis of Stilboestrol-Induced Renal Tumours in the Male Golden Hamster', *Brit. Jour. Cancer*, Vol. 8 (1954), pp. 451–7.

PAGE 208 Kirkman, Hadley, *Estrogen-Induced Tumors of the Kidney in the Syrian Hamster*. U.S. Public Health Service, Natl Cancer Inst. Monograph No. 1 (December 1959).

PAGE 208 Ayre, J. E. and Bauld, W. A. G., 'Thiamine Deficiency and High Estrogen Findings in Uterine Cancer and in Menorrhagia', *Science*, Vol. 103, No. 2676 (12 April 1946), pp. 441–5.

PAGES 208–9 Rhoads, C. P., 'Physiological Aspects of Vitamin Deficiency', *Proc. Inst. Med. Chicago*, Vol. 13 (1940), p. 198.

PAGES 208–9 Sugiura, K. and Rhoads, C. P., 'Experimental Liver Cancer in Rats and Its Inhibition by Rice-Bran Extract, Yeast, and Yeast Extract', *Cancer Research*, Vol. 1 (1941), pp. 3–16.

PAGE 209 Martin, H., 'The Precancerous Mouth Lesions of Avitaminosis B. Their Etiology, Response to Therapy and

Relationship to Intraoral Cancer', *Am. Jour. Surgery*, Vol. 57 (1942), pp. 195–225.

PAGE 209   Tannenbaum, A., 'Nutrition and Cancer', in Freddy Homburger, ed., *Physiopathology of Cancer*, New York: Harper, 1959, 2nd ed. A. Paul B. Hoeber Book, p. 552; London, Cassell, 1959.

PAGE 209   Symeonidis, A., 'Post-starvation Gynecomastia and Its Relationship to Breast Cancer in Man', *Jour. Natl. Cancer Inst.*, Vol. 11 (1950), p. 656.

PAGES 208–9   Davies, J. N. P., 'Sex Hormone Upset in Africans', *Brit. Med. Jour.*, Vol. 2 (1949), pp. 676–9.

PAGE 209   Hueper, W. C., 'Potential Role of Non-Nutritive Food Additives'.

PAGE 209   VanEsch, G. J., *et al.*, 'Production of Skin Tumours in Mice by Carbamates'.

PAGE 210   Berenblum, I. and Trainin, N., 'Possible Two-Stage Mechanism in Experimental Leukemogenesis', *Science*, Vol. 132 (1 July 1960), pp. 40–1.

PAGE 210   Hueper, W. C., 'Cancer Hazards from Natural and Artificial Water Pollutants', *Proc.*, Conf. on Physiol. Aspects of Water Quality, Washington, D.C., 8–9 September 1960, pp. 181–93. U.S. Public Health Service.

PAGE 210   Hueper, W. C. and Payne, W. W., 'Observations on Occurrence of Hepatomas in Rainbow Trout.'

PAGES 211–13   Hueper, W. C., To author.

PAGE 212   Sloan-Kettering Inst. for Cancer Research, *Biennial Report*, 1957–9.

CHAPTER 15: NATURE FIGHTS BACK

PAGE 214   Briejèr, C. J., 'The Growing Resistance of Insects to Insecticides', *Atlantic Naturalist*, Vol. 13 (1958), No. 3, pp. 149–55.

PAGE 215   Metcalf, Robert L., 'The Impact of the Development of Organophosphorus Insecticides upon Basic and Applied Science', *Bull. Entomol. Soc. Am.*, Vol. 5 (March 1959), pp. 3–15.

PAGE 216   Ripper, W. E., 'Effect of Pesticides on Balance of

Arthropod Populations', *Annual Rev. Entomol.*, Vol. 1 (1956), pp. 403–38.

PAGE 216 Allen, Durward L., *Our Wildlife Legacy*, New York, Funk & Wagnalls, 1954, pp. 234–6; London, Mayflower, 1954.

PAGE 217 Sabrosky, Curtis W., 'How Many Insects Are there?' *Yearbook of Agric.*, U.S. Dept of Agric., 1952, pp. 1–7.

PAGE 217 Bishopp, F. C., 'Insect Friends of Man', *Yearbook of Agric.*, U.S. Dept of Agric., 1952, pp. 79–87.

PAGE 217 Klots, Alexander B. and Klots, Elsie B., 'Beneficial Bees, Wasps, and Ants', *Handbook on Biological Control of Plant Pests*, pp. 44–6, Brooklyn Botanic Garden. Reprinted from *Plants and Gardens*, Vol. 16 (1960), No. 3.

PAGE 218 Hagen, Kenneth S., 'Biological Control with Lady Beetles', *Handbook on Biological Control of Plant Pests*, pp. 28–35.

PAGE 218 Schlinger, Evert I., 'Natural Enemies of Aphids', *Handbook on Biological Control of Plant Pests*, pp. 36–42.

PAGE 219 Bishopp, F. C., 'Insect Friends of Man'.

PAGE 220 Ripper, W. E., 'Effect of Pesticides on Arthropod Populations'.

PAGE 220 Davies, D. M., 'A Study of the Black-fly Population of a Stream in Algonquin Park, Ontario', *Transactions*, Royal Canadian Inst., Vol. 59 (1950), pp. 121–59.

PAGE 220 Ripper, W. E., 'Effect of Pesticides on Arthropod Populations'.

PAGE 220 Johnson, Philip C., *Spruce Spider Mite Infestations in Northern Rocky Mountain Douglas-Fir Forests*. Research Paper 55, Inter-mountain Forest and Range Exper. Station, U.S. Forest Service, Ogden, Utah, 1958.

PAGE 221 David, Donald W., 'Some Effects on DDT on Spider Mites', *Jour. Econ. Entomol.*, Vol. 45 (1952), No. 6, pp. 1011–19.

PAGE 221 Gould, E. and Hamstead, E. O., 'Control of the Red-banded Leaf Roller', *Jour. Econ. Entomol.*, Vol. 41 (1948), pp. 887–90.

PAGE 222 Pickett, A. D., 'A Critique on Insect Chemical Control Methods', *Canadian Entomologist*, Vol. 81 (1949), No. 3, pp. 1–10.

PAGE 222   Joyce, R. J. V., 'Large-scale Spraying of Cotton in the Gash Delta in Eastern Sudan', *Bull. Entomol. Research*, Vol. 47 (1956), pp. 390–413.

PAGE 223   Long, W. H., *et al.*, 'Fire Ant Eradication Program Increases Damage by the Sugarcane Borer', *Sugar Bull.*, Vol. 37 (1958), No. 5, pp. 62–3.

PAGE 223   Luckmann, William H., 'Increase of European Corn Borers Following Soil Application of Large Amounts of Dieldrin', *Jour. Econ. Entomol.*, Vol. 53 (1960), No. 4, pp. 582–4.

PAGE 223   Haeussler, G. J., 'Losses Caused by Insects', *Yearbook of Agric.*, U.S. Dept of Agric., 1952, pp. 141–6.

PAGE 223   Clausen, C. P., 'Parasites and Predators', *Yearbook of Agric.*, U.S. Dept of Agric., 1952, pp. 380–8.

PAGE 223   Clausen, C. P., *Biological Control of Insect Pests in the Continental United States*, U.S. Dept of Agric. Technical Bulletin No. 1139 (June 1956), pp. 1–151.

PAGE 224   De Bach, Paul, 'Application of Ecological Information to Control of Citrus Pests in California', *Proc.*, 10th Internatl Congress of Entomologists (1956), Vol. 3 (1958), pp. 187–94.

PAGE 224   Laird, Marshall, 'Biological Solutions to Problems Arising from the Use of Modern Insecticides in the Field of Public Health', *Acta Tropica*, Vol. 16 (1959), No. 4, pp. 331–55.

PAGE 224   Harrington, R. W. and Bidlingmayer, W. L., 'Effects of Dieldrin on Fishes and Invertebrates of a Salt Marsh', *Jour. Wildlife Management*, Vol. 22 (1958), No. 1, pp. 76–82.

PAGE 225   *Liver Flukes in Cattle*. U.S. Dept of Agric. Leaflet No. 493 (1961).

PAGE 225   Fisher, Theodore W., 'What is Biological Control?' *Handbook on Biological Control of Plant Pests*, pp. 6–18, Brooklyn Botanic Garden. Reprinted from *Plants and Gardens*, Vol. 16 (1960), No. 3.

PAGE 226   Jacob, F. H., 'Some Modern Problems in Pest Control', *Science Progress*, No. 181 (1958), pp. 30–45.

PAGE 226   Pickett, A. D. and Patterson, N. A., 'The Influence of Spray Programs on the Fauna of Apple Orchards in Nova

Scotia. IV. A Review', *Canadian Entomologist*, Vol. 85 (1953), No. 12, pp. 472-8.

PAGE 226 Pickett, A. D., 'Controlling Orchard Insects', *Agric. Inst. Rev.*, March–April 1953.

PAGE 226 Pickett, A. D., 'The Philosophy of Orchard Insect Control', *79th Annual Report*, Entomol. Soc. of Ontario (1948), pp. 1-5.

PAGE 227 Pickett, A. D., 'The Control of Apple Insects in Nova Scotia'. Mimeo.

PAGE 228 Ullyett, G. C., 'Insects Man and the Environment', *Jour. Econ. Entomol.*, Vol. 44 (1951), No. 4, pp. 459-64.

CHAPTER 16: THE RUMBLINGS OF AN AVALANCHE

PAGE 229 Babers, Frank H., *Development of Insect Resistance to Insecticides*. U.S. Dept. of Agric., E 776 (May 1949).

PAGE 229 Babers, Frank H. and Pratt, J. J., *Development of Insect Resistance to Insecticides. II. A Critical Review of the Literature up to 1951*. U.S. Dept of Agric., E 818 (May 1951).

PAGE 230 Brown, A. W. A., 'The Challenge of Insecticide Resistance', *Bull. Entomol. Soc. Am.*, Vol. 7 (1961), No. 1, pp. 6-19.

PAGE 230 Brown, A. W. A., 'Development and Mechanism of Insect Resistance to Available Toxicants', *Soap and Chem. Specialties*, January 1960.

PAGE 230 *Insect Resistance and Vector Control*. World Health Organ. Technical Report Ser. No. 153 (Geneva, 1958), p. 5.

PAGE 230 Elton, Charles S., *The Ecology of Invasions by Animals and Plants*, New York, Wiley, 1958, p. 181; London, Methuen, 1958.

PAGE 230 Babers, Frank H. and Pratt, J. J., *Development of Insect Resistance to Insecticides*, II.

PAGES 231-2 Brown, A. W. A., *Insecticide Resistance in Arthropods*. World Health Organ. Monograph Ser. No. 38 (1958), pp. 13, 11.

PAGE 232 Quarterman, K. D. and Schoof, H. F., 'The Status of Insecticide Resistance in Arthropods of Public Health

Importance in 1956', *Am. Jour. Trop. Med. and Hygiene*, Vol. 7 (1958), No. 1, pp. 74–83.

PAGE 232 Brown, A. W. A., *Insecticide Resistance in Arthropods*.

PAGE 232 Hess, Archie D., 'The Significance of Insecticide Resistance in Vector Control Programs', *Am. Jour. Trop. Med. and Hygiene*, Vol. 1 (1952), No. 3, pp. 371–88.

PAGE 233 Lindsay, Dale R. and Scudder, H. I., 'Nonbiting Flies and Disease', *Annual Rev. Entomol.*, Vol. 1 (1956), pp. 323–46.

PAGE 233 Schoof, H. F. and Kilpatrick, J. W., 'House Fly Resistance to Organo-phosphorus Compounds in Arizona and Georgia', *Jour. Econ. Entomol.*, Vol. 51 (1958), No. 4, 546.

PAGE 233 Brown, A. W. A., 'Development and Mechanism of Insect Resistance'.

PAGE 233 Brown, A. W. A., *Insecticide Resistance in Arthropods*.

PAGE 234 Brown, A. W. A., 'Challenge of Insecticide Resistance'.

PAGE 234 Brown, A. W. A., *Insecticide Resistance in Arthropods*.

PAGE 235 Brown, A. W. A., 'Development and Mechanism of Insect Resistance'.

PAGE 235 Brown, A. W. A., *Insecticide Resistance in Arthropods*.

PAGE 235 Brown, A. W. A., 'Challenge of Insecticide Resistance'.

PAGE 235 *New York Herald Tribune*, 22 June 1959; also Pallister, J. C., To author, 6 November 1959.

PAGES 235–6 Anon., 'Brown Dog Tick Develops Resistance to Chlordane', *New Jersey Agric.*, Vol. 37 (1955), No. 6, pp. 15–16.

PAGE 236 Brown, A. W. A., 'Challenge of Insecticide Resistance'.

PAGE 236 Hoffmann, C. H., 'Insect Resistance', *Soap*, Vol. 32 (1956), No. 8, pp. 129–32.

PAGE 237 Brown, A. W. A., *Insect Control by Chemicals*, New York, Wiley, 1951; London, Chapman & Hall, 1951.

PAGE 237 Briejèr, C. J., 'The Growing Resistance of Insects to Insecticides', *Atlantic Naturalist*, Vol. 13 (1958), No. 3, pp. 149–55.

PAGE 237 Laird, Marshall, 'Biological Solutions to Problems Arising from the Use of Modern Insecticides in the Field of Public Health', *Acta Tropica*, Vol. 16 (1959), No. 4, pp. 331–55.

PAGE 237 Brown, A. W. A., *Insecticide Resistance in Arthropods*.

PAGE 238 Brown, A. W. A., 'Development and Mechanism of Insect Resistance'.

PAGE 239 Briejèr, C. J., 'Growing Resistance of Insects to Insecticides'.

PAGE 239 'Pesticides – 1959', *Jour. Agric. and Food Chem.*, Vol. 7 (1959), No. 10, p. 680.

PAGE 239 Briejèr, C. J., 'Growing Resistance of Insects to Insecticides'.

CHAPTER 17: THE OTHER ROAD

PAGE 240 Swanson, Carl P., *Cytology and Cytogenetics*, Englewood Cliffs, N.J., Prentice-Hall, 1957.

PAGE 241 Knipling, E. F., 'Control of Screw-Worm Fly by Atomic Radiation', *Sci. Monthly*, Vol. 85 (1956), No. 4, pp. 195–202.

PAGE 241 Knipling, E. F., *Screwworm Eradication: Concepts and Research Leading to the Sterile-Male Method*. Smithsonian Inst. Annual Report, Publ. 4365 (1959).

PAGE 241 Bushland, R. C., *et al.*, 'Eradication of the Screw-Worm Fly by Releasing Gamma-Ray-Sterilized Males among the Natural Population', *Proc.*, Internatl Conf. on Peaceful Uses of Atomic Energy, Geneva, August 1955, Vol. 12, pp. 216–20.

PAGE 242 Lindquist, Arthur W., 'The Use of Gamma Radiation for Control or Eradication of the Screwworm', *Jour. Econ. Entomol.*, Vol. 48 (1955), No. 4, pp. 467–9.

PAGE 242 Lindquist, Arthur W., 'Research on the Use of Sexually Sterile Males for Eradication of Screw-Worms', *Proc.*, Inter-Am. Symposium on Peaceful Applications of Nuclear Energy, Buenos Aires, June 1959, pp. 229–39.

PAGE 243 'Screwworm vs. Screwworm', *Agric. Research*, July 1958, p. 8. U.S. Dept of Agric.

PAGE 243 'Traps Indicate Screwworm May Still Exist in South-east', U.S. Dept of Agric. Release No. 1502–59 (3 June 1959). Mimeo.

PAGE 244 Potts, W. H., 'Irradiation and the Control of Insect Pests', *Times* (London) Sci. Rev., Summer 1958, pp. 13–14.

PAGE 244 Knipling, E. F., *Screwworm Eradication: Sterile-Male Method*.

PAGE 244 Lindquist, Arthur W., 'Entomological Uses of Radioisotopes', in *Radiation Biology and Medicine*. U.S. Atomic Energy Commission, 1958. Ch. 27, Pt 8, pp. 688–710.

PAGE 244 Lindquist, Arthur W., 'Research on the Use of Sexually Sterile Males'.

PAGE 244 'USDA May Have New Way to Control Insect Pests with Chemical Sterilants', U.S. Dept of Agric. Release No. 3587–61 (1 November 1961). Mimeo.

PAGE 245 Lindquist, Arthur W., 'Chemicals to Sterilize Insects', *Jour. Washington Acad. Sci.*, November 1961, pp. 109–14.

PAGE 245 Lindquist, Arthur W., 'New Ways to Control Insects', *Pest Control Mag.*, June 1961.

PAGE 245 La Brecque, G. C., 'Studies with Three Alkylating Agents As House Fly Sterilants', *Jour. Econ. Entomol.*, Vol. 54 (1961), No. 4, pp. 684–9.

PAGE 245 Knipling, E. F., 'Potentialities and Progress in the Development of Chemosterilants for Insect Control', paper presented at Annual Meeting Entomol. Soc. of Am., Miami, 1961.

PAGE 245 'Use of Insects for Their Own Destruction', *Jour. Econ. Entomol.*, Vol. 53 (1960), No. 3, pp. 415–20.

PAGE 246 Mitlin, Norman, 'Chemical Sterility and the Nucleic Acids', paper presented 27 November 1961, Symposium on Chemical Sterility, Entomol. Soc. of Am., Miami.

PAGE 246 Alexander, Peter, To author, 19 February 1962.

PAGE 246 Eisner, T., 'The Effectiveness of Arthropod Defensive Secretions', in Symposium 4 on 'Chemical Defensive Mechanisms', 11th Internatl Congress of Entomologists, Vienna (1960), pp. 264–7. Offprint.

PAGE 246 Eisner, T., 'The Protective Role of the Spray Mechanisms of the Bombardier Beetle, *Brachynus ballistarius* Lec.', *Jour. Insect Physiol.*, Vol. 2 (1958), No. 3, pp. 215–20.

PAGE 246 Eisner, T., 'Spray Mechanism of the Cockroach *Diploptera punctata*', *Science*, Vol. 128, No. 3316 (18 July 1958), pp. 148–9.

PAGE 246–7 Williams, Carroll M., 'The Juvenile Hormone', *Sci. American*, Vol. 198, No. 2 (February 1958), p. 67.

PAGE 247 '1957 Gypsy-Moth Eradication Program', U.S. Dept of Agric. Release 858–57–3. Mimeo.

PAGES 247–8 Jacobson, Martin, *et al.*, 'Isolation, Identification, and Synthesis of the Sex Attractant of Gypsy Moth', *Science*, Vol. 132, No. 3433 (14 October 1960), p. 1011.

PAGE 248 Brown, William L., Jr, 'Mass Insect Control Programs: Four Case Histories', *Psyche*, Vol. 68 (1961), Nos. 2–3, pp. 75–111.

PAGE 248 Christenson, L. D., 'Recent Progress in the Development of Procedures for Eradicating or controlling Tropical Fruit Flies', *Proc.*, 10th Internatl Congress of Entomologists (1956), Vol. 3 (1958), pp. 11–16.

PAGE 248 Hoffmann, C. H., 'New Concepts in Controlling Farm Insects', address to Internatl Assn Ice Cream Manuf. Conv., 27 October 1961. Mimeo.

PAGE 249 Frings, Hubert and Frings, Mabel, 'Uses of Sounds by Insects', *Annual Rev. Entomol.*, Vol. 3 (1958), pp. 87–106.

PAGE 249 *Research Report*, 1956–1959. Entomol. Research Inst. for Biol. Control, Belleville, Ontario, pp. 9–45.

PAGE 249 Kahn, M. C. and Offenhauser, W., Jr, 'The First Field Tests of Recorded Mosquito Sounds Used for Mosquito Destruction', *Am. Jour. Trop. Med.*, Vol. 29 (1949), pp. 800–27.

PAGE 249 Wishart, George, To author, 10 August 1961.

PAGE 249 Beirne, Bryan, To author, 7 February 1962.

PAGE 249 Frings, Hubert, To author, 12 February 1962.

PAGE 249 Wishart, George, To author, 10 August 1961.

PAGE 249 Frings, Hubert, *et al.*, 'The Physical Effects of High Intensity Air-Borne Ultrasonic Waves on Animals', *Jour. Cellular and Compar. Physiol.*, Vol. 31 (1948), No. 3, pp. 339–58.

PAGE 249   Steinhaus, Edward A., 'Microbial Control – The Emergence of an Idea', *Hilgardia*, Vol. 26, No. 2 (October 1956), pp. 107–60.

PAGE 250   Steinhaus, Edward A., 'Concerning the Harmlessness of Insect Pathogens and the Standardization of Microbial Control Products', *Jour. Econ. Entomol.*, Vol. 50, No. 6 (December 1957), pp. 715–20.

PAGE 250   Steinhaus, Edward A., 'Living Insecticides', *Sci. American*, Vol. 195, No. 2 (August 1956), pp. 96–104.

PAGE 250   Angus, T. A. and Heimpel, A. E., 'Microbial Insecticides', *Research for Farmers*, Spring 1959, pp. 12–13. Canada Dept. of Agric.

PAGE 250   Heimpel, A. M. and Angus, T. A., 'Bacterial Insecticides', *Bacteriol. Rev.*, Vol. 24 (1960), No. 3, pp. 266–8.

PAGE 251   Briggs, John D., 'Pathogens for the Control of Pests', *Biol. and Chem. Control of Plant and Animal Pests.* Washington, D.C., Am. Assn Advancement Sci., 1960, pp. 137–48.

PAGE 251   'Tests of a Microbial Insecticide against Forest Defoliators', *Bi-Monthly Progress Report*, Canada Dept of Forestry, Vol. 17, No. 3 (May–June 1961).

PAGE 251   Steinhaus, Edward A., 'Living Insecticides'.

PAGE 251   Tanada, Y., 'Microbial Control of Insect Pests', *Annual Rev. Entomol.*, Vol. 4 (1959), pp. 277–302.

PAGES 251–2   Steinhaus, Edward A., 'Concerning the Harmlessness of Insect Pathogens'.

PAGE 252   Clausen, C. P., *Biological Control of Insect Pests in the Continental United States.* U.S. Dept of Agric. Technical Bulletin No. 1139 (June 1956), pp. 1–151.

PAGE 252   Hoffmann, C. H., 'Biological Control of Noxious Insects, Weeds', *Agric. Chemicals*, March–April 1959.

PAGE 252   De Bach, Paul, 'Biological Control of Insect Pests and Weeds', *Jour. Applied Nutrition*, Vol. 12 (1959), No. 3, pp. 120–34.

PAGE 253   Ruppertshofen, Heinz, 'Forest-Hygiene', address to 5th World Forestry Congress, Seattle, Wash. (29 August–10 September 1960).

PAGE 254   Ruppertshofen, Heinz, To author, 25 February 1962.

PAGE 254   Gösswald, Karl, *Die Rote Waldameise im Dienste der Waldhygiene*, Lüneburg, Metta Kinau Verlag, n.d.

PAGE 254   Gösswald, Karl, To author, 27 February 1962.

PAGE 255   Balch, R. E., 'Control of Forest Insects', *Annual Rev. Entomol.*, Vol. 3 (1958), pp. 449–68.

PAGE 255   Buckner, C. H., 'Mammalian Predators of the Larch Sawfly in Eastern Manitoba', *Proc.*, 10th Internatl Congress of Entomologists (1956), Vol. 4 (1958), pp. 353–61.

PAGE 255   Morris, R. F., 'Differentiation by Small Mammal Predators between Sound and Empty Cocoons of the European Spruce Sawfly', *Canadian Entomologist*, Vol. 81 (1949), No. 5.

PAGE 256   MacLeod, C. F., 'The Introduction of the Masked Shrew into Newfoundland', *Bi-Monthly Progress Report*, Canada Dept of Agric., Vol. 16, No. 2 (March–April 1960).

PAGE 256   MacLeod, C. F., To author, 12 February 1962.

PAGE 256   Carroll, W. J., To author, 8 March 1962.

Index

Acetylcholine, 42–3

Adipose tissue, storage of chemicals in, 170–1

ADP (adenosine diphosphate), 180, 181

Africa, cancer in tribes of, 208; results of DDT spraying in, 222

Agriculture Department. See U.S. Department of Agriculture

Alabama, fire ants in, 100, 101, 149, 150, 156

Alabama Co-operative Wildlife Research Unit, 150, 152

Aldrin, 39, 40; nitrification affected by, 65; persistence in soil, 65; used against Japanese beetle in Michigan, 88–9, 90–91; birds killed by, 91, 92; toxicity, 95; as seed coating, 119

Alexander, Dr Peter, 188, 246

Alfalfa caterpillar, virus used against, 251

American Cancer Society, 195

American Medical Association, 159, 170

American Society of Ichthyologists and Herpetologists, 132

Aminotriazole, 166; carcinogenic nature of, 149, 198–9

Amitrol. See Aminotriazole

Anaemia, aplastic, 200, 201

Anopheles: mosquitoes, malaria carried by, 224, 231; resistant to DDT, 234

Anoxia, caused by nitrates, 81; consequences of, 182

Ant, fire, 147–54, 156–7, 222; forest red, as insect predators, 254

Antelope, pronghorn, 70, 72

Appleworm. See Codling moth

Arant, Dr F. S., 149

Army Chemical Corps, Rocky Mountain Arsenal of, 53

Arsenic, 32–33; in herbicides, 48; as carcinogen, 60, 196–7; soil poisoned by, 66; cows killed by, 75; in crabgrass killers, 83–4; human exposure to, 209

ATP (adenosine triphosphate), 180–81, 183, 185, 203–4

Attractants, insect sex, 247

Audubon Society, Detroit, 91; Michigan, 91; National, 101, 137; Florida, 113

Auerbach, Charlotte, 186

Austin, Texas, fish killed by chemicals near, 135

B vitamins, 208–9

Bacillus thuringiensis, 250

Bacterial insecticides, 249–52. See also Milky disease

Baker, Dr Maurice F., 150

Balance of nature, 215–16

Bantu tribes, cancer in, 208

Barker, Dr Roy, 103, 104

Baton Rouge, birds killed by insecticides in, 101

Beaver, 72, 73

Beck, Professor Herbert H., 114–
115

Bedbugs, 238

Beekeeping, 33, 146

Bees, effect of parathion on, 43;
dependence on 'weeds', 77;
killed by insecticides, 146;
deaths from sting of, 149

Beetle, used in weed control, 85;
Japanese, 90–96, 222–3, 252;
white-fringed, 150; vedalia,
223–4, 252

Benson, Ezra, 150

Bent, Arthur C., *Life Histories*,
108

Benzene, leukaemia caused by,
206

BHC (benzene hexachloride),
effect on nitrification, 64, 65;
persistence in soil, 65; sweet
potatoes and peanuts con-
taminated by, 66–67; as its
isomer, lindane, 175; plant
mutations caused by, 189; and
blood disorders, 201, 202, 206;
arthropods resistant to, 231,
233–4

Bernard, Richard F., 115

Bidlingmayer, W. L., 137

Biesele, John J., 205

Bingham, Millicent Todd, 74

Biocides, 25

Biological control of insects, 224,
225–8, 240–57

Birds, fish-eating, killed by insecticides, 55–8; reproduction
affected adversely by herbicides, 80; killed by herbicides,
84; killed by aldrin, 91, 92, 99,
119; killed by dieldrin, 93–4;
killed by elm spraying, 100–10;
apparent sterility in (eagles),
113, 114; killed by seed treatment in England, United
States, 117–20; killed by fire
ant spraying programme, 151–
153; encouragement of, in
modern forests, 253–4. *See also*
Sterility, and various names of
birds, such as Eagles, Grebes,
Grouse, Gulls, Robins, Warblers

Blindness, in fish, caused by
DDT, 127

Blood disorders, insecticides and,
199–203

Blue Island, Illinois, 92

Bobwhite quail, 152

Bollworm, 222

Bone marrow, chemicals with
affinity for, 206

Bonin Islands, 248

Boyes, Mrs Ann, 91

Bridger National Forest, 72–
73

Briejèr, C. J., 82, 214, 237,
239

British Columbia, forest spraying
injures salmon in, 129

British Trust for Ornithology,
117

Broley, Charles, 113, 116

Brooks, Professor Maurice, 101

Broun, Maurice, 114

Brown, Dr A. W. A., 231, 232,
236

'Brush control' spraying, 73–6;
selective, 78–9

Budworm, black-headed, DDT
spraying for in British Columbia, 129

Budworm, spruce, DDT spraying for in eastern Canada, 123–125; in Maine, 127; in Montana, 128; use of microbial disease against, 251

Burnet, Sir F. Macfarlane, 188, 206

Butler, Dr Philip, 141

Cactus, insect enemy used to control, 85–6

California Citrus Experiment Station, 230

California Department of Public Health, 58

Canada, spraying programmes in, 129; 'forest hygiene' programmes in, 255–6

Cancer: hazards from polluted water, 60; and cellular oxidation, 182; natural causative agents, 193; and man-made carcinogens, 193–4; and industrial carcinogens, 194; increase in, 195; in children, 195; and pesticides as carcinogens, 196–203, 209; Warburg theory of origin, 203–5; and chromosome abnormality, 205–6; urethane as cause of, 207; possible indirect causes, 207–9; and imbalance of sex hormones, 207–209; protective role of vitamins against, 208–9; multiple exposure to causative agents of, 209–12; search for cause *vs.* search for cure, 212–13. *See also* Leukaemia

Carbamates, 189, 207

Carbon tetrachloride, molecular structure, 35

Carcinogens, 193–4; industrial, 194–5, 199; pesticides as, 196–198, 199–203; herbicides as, 198–9

Carroll, Lewis, 167

Carrots, insecticides absorbed by, 66

Cats, affected by aldrin, 91; dieldrin fatal to, 94

Cattle: killed by arsenical insecticides, 75; attracted to and killed by plants sprayed with 2,4-D, 80–81; killed by fire ant programme, 152–3

Cell division, 186–7; and cancer, 203–5

Cellular oxidation, 178–81; effect of insecticides upon, 181–4

Chaoborus astictopus, gnat, 56

Chemicals, general, new to human environment, 24; insect-killing, new, 24–5; insecticidal, growth of production of, 32; biological potency of, 32; dangerous interaction of, 45, 209; recurrent exposure to, 158; less toxic, 167; stored in human body, 170; parallel between radiation and, 185. *See also* Herbicides, Insecticides, Pesticides, and various chemicals by name

Chester Beatty Research Institute (London), 246

Chickadees, 108

Chlordane, 36, 38–39; persistence in soil, 65; in crabgrass killers, 84; toxic to fish, 130, 135; household use questionable, 159; and blood disorders, 201, 202; arthropods resistant to, 232; roaches and ticks resistant to, 235–6

Chloroform, molecular structure, 34

Cholera epidemic, London, 212

Cholinesterase, 42, 174

Chromosomes: and mitosis, 186–187; effect of environmental factors on, 188; effect of pesticides on, 189–91; abnormality of, in chronic leukaemia, 190–192; abnormality of, and birth defects, 191–2; abnormality of, and cancer, 205–6

Cigarettes, arsenic content of, 66

CIPC, 198, 207, 209

Cirrhosis, increase of, 172

Citrus industry, scale insect a threat to, 223–4, 252

Clams, 140

Clear Lake, California, 56–9

Cockroaches, 236

Codling moth, in Nova Scotia, 221; resistant to sprays, 230; resistant to DDT, 236

Colorado River, fish destruction in, 135–6

Commercial Fisheries, Bureau of, 140, 141

Congenital defects, due to anoxia, 182; due to chromosome damage, 191–2

Connecticut Arboretum, 75

Corn borer, 222–3, 251

Cornell University, 246; Agricultural Experimental Station, 146

Cottam, Dr Clarence, 152–3

Coyotes, 217

Crabgrass, 83–4, 161–2

Crabs, dieldrin fatal to, 138–9

Cranberry-weed killer, 49, 166, 198–9

Cranbrook Institute of Science, 105, 109

Culex mosquitoes, 232

Curaçao, eradication of screw-worm on, 242–3

Czechoslovakia, biological warfare experiments in, 251

Darwin, Charles, *The Formation of Vegetable Mould*, 63

Darwin, Erasmus, 252

Davis, Professor David E., 190

DDD, 55; used against gnats at Clear Lake, 56–59; physiological effect of, 58–59

DDT (dichloro-diphenyl-trichloro-ethane), discovery, 35; stored in human body, 36, 161, 162, 163; passed from one organism to another, 37; used against spruce budworm, 52, 123–6; persistence in soil, 65; birds poisoned by, 100, 104, 106, 108, 115–16, 119–20; used for Dutch elm disease, 104–5; effect on reproduction of birds, 105, 115–16, 184, 185; stored in tissues of fish, 128; toxic to fish, 134, 135; aerial spraying of, 144–6; in milk, 145; in leaf crops, 146; effect on nervous system, 172–3; as uncoupler, 182; genetic effects on mosquitoes, 189; as carcinogen, 198; and blood disorders, 199, 200, 201; human exposure to, 209; certain insects increase under spraying, 220, 221–2, 226–7; effect on spider mite, 221; used against typhus, 232; flies develop resistance to, 233; mosquitoes resistant to, 234–5, 237; agricultural insects resistant to, 236

DeBach, Dr Paul, 224, 252

Deer, mule, 71, 72; Kaibab, 216–217

Defects, congenital. *See* Congenital defects

Denmark, flies become resistant in, 232–3

Detergents, indirect role in carcinogenesis, 209–11

Detroit, spraying for Japanese beetle in, 88, 90–91

Detroit Audubon Society, 91

Detroit *News*, 90

De Witt, Dr James, 115–16

Dieldrin, 36, 39–40; aldrin converted to, in soil, 65; effects of spraying with, in Sheldon, Ill., 93–95; toxicity, 93; cats killed by, 94; toxic to fish, 130, 131; toxic to shrimp, 140; used against fire ants, 150; ruled unsuitable in forage, 154; delayed effects on nervous system, 175; flies resistant to, 333; banana root borer resistant to, 250–51

Diels, Otto, 39

'Dinitro' herbicides, 48

Dinitrophenol, 48–9, 182, 185

Disease, environmental, 168–77; insect-borne, 231; as weapon against insects, 249–52

Douglas, Justice William O., 72, 73, 76, 145

Dragonflies, 218

Dubos, Dr René, 169

Dutch elm disease, 102; spraying for, 102, 110; controlled by sanitation, 110–12

Dutch Plant Protection Service, 82

Eagles, insecticides a threat to, 113–14, 116

Earthworms, Darwin on, 63; poisoned by spraying, 103–4, 106, 141

East Lansing, Mich., robin population affected by spraying at, 102–5

Ecology, 169

Ecology of Invasions, The (Elton), 27

Egler, Dr Frank, 78

Egypt, flies develop resistance in, 233

Eliassen, Professor Rolf, 51

Elm: American, and Dutch elm disease, 27, 102, 110; European, 112

Elton, Dr Charles, 27, 28, 112, 230

Endrin, 39, 40–41; toxic to fish, 130, 131, toxic to shrimp, 140

England, use of arsenical weed killers in, 48; birds affected by seed treatment in, 117–19

Entomologists, chemical control favoured by some, 225–6

Environment, adjustment of life to, 24; man's contamination of, 26–30

Enzymes, function, 32, 182; affected by organic phosphates, 42–43; cholinesterase, 42–3, 174, 176; liver, 45, 171; role in oxidation, 179, 180, 182; in flies, 238

Eskimos, DDT in fat of, 163

Farm surpluses and insect control, 26

Fawks, Elton, 114

Federal Aviation Agency, 90

Field Notes, Audubon, 101

Fire ant, programme against, 147–53, 156–7, 222; effective method of control, 157

'Fire damp', 34

Fish, killed by insecticides, 51–52, 116, 122–38, 139–40; affected by herbicides, 72, 73; blinded by DDT, 127

Fish and Wildlife Service. *See* U.S. Fish and Wildlife Service

Fisheries Research Board of Canada, 124

'Flareback', insects', after spraying, 25, 220–24

Flint Creek, Alabama, 133

Florida, fish destruction in, 132; pesticide pollution in salt marshes in, 137–9; abandons broad fire ant control programme, 156; mosquitoes become resistant in, 235

Flukes, blood and liver, 225

Fly, fruit, 186, 248; screw-worm, 242–3; Hessian, 248; melon, 248. *See also* Housefly

Food, chemical residues in, 162–167; contamination in warehouses, 164. *See also* Milk

Food and Drug Administration. *See* U.S. Food and Drug Administration

'Forest hygiene', 253

Forest Service. *See* U.S. Forest Service

France, birds affected by insecticides in, 116–17

Freiberg, Germany, arsenic-contamination affects animals at, 197

Frings, Hubert and Mable, 249

Game Birds Association (British), 117

Gardening, poisons used in, 160–162

Genelly, Dr Richard, 115

Genes, 185–7

Genetic effect, of chemicals, 25, 185, 186; of radiation, 186

'Ginger paralysis', 176

Gnat, *Chaoborus astictopus*, 56

Goatweed. *See* Klamath weed

Gösswald, Professor Karl, 254

Grebes, western, 55, 57

Gromme, Owen J., 108–9

Groundwater, contamination of, 52–3, 59–60

Grouse, sage, 70, 72

Gulls, 55; California, DDD residues in, 58; laughing, affected by spraying of marshes, 138

Gynandromorphs, 189

'Gyplure', 247

Gypsy moth, 143; importation of natural enemies of, 143; aerial spraying for, 144–7; secretion as weapon against, 247–8; synthetic lure isolated, 247

Hargraves, Dr Malcolm, 200, 201, 202

Harrington, R. W., Jr, 137

Hawk Mountain Sanctuary, 114

Hayes, Dr Wayland, Jr, 37

Health problems, new environmental, 168–77

Hepatitis, 39; increase of, 172

Heptachlor, 39; effect on nitrification, 64–5; persistence in soil, 65; effect on hops sprayed with, 67–8; effect on wildlife, Joliet, Illinois, 92; toxic to fish, 130, 131; used against fire ants, 150, 151, 152, 153, 154, 156; ruled unsuitable on forage, 154; peculiar nature of, 155; use results in increase of sugarcane borer, 222

Herbicides, toxic effects of, 47–

49, 80; used against sagebrush, 69–72; used for roadside 'brush control', 73–9; animals attracted to plants sprayed with, 80; possible effects on reproduction in birds, 80; toxic to plankton, 140; as agents of chromosome damage, 189–90; as carcinogens, 198

Hessian fly, 248

Hickey, Professor Joseph, 105

Hinsdale, Illinois, birds killed by DDT in, 100

Hiroshima, leukaemia among survivors of, 199

Hops, destroyed by heptachlor, 67–8

Hormones, sex, imbalance of, and cancer development, 207–9

Housefly, diseases carried by, 231; resistance to DDT and other chemicals, 232–3, 238; pilot projects in sterilization of, 244

Hueper, Dr W. C., on arsenicals, 33; on contaminated drinking water, 60; on congenital and infant cancer, 195–6, 207; *Occupational Tumours*, 196, 197; on DDT as carcinogen, 198; on epidemic of cancer in trout, 211; on eliminating causative agents of cancer, 212–13

Hurricane: Edna (1954), 125; of 1938, 143

Huxley, Thomas, 216

Hydrocarbons, chlorinated, 33–41; storage of, 36, 39, 40, 171; persistence in soil, 65; sensitivity of fish to, 130; in food crops, 163–7; effect on liver, 171–2, 174, 208; effect on nervous system, 172–7; genetic effects of, 190

Illinois Agriculture Department, 92

Illinois Natural History Survey, 93, 94, 103; report quoted, 94–5

Industry, malignancies traceable to, 194–5, 199

Insecticides: abuses in use, general, 29; botanical, 32, 167; synthetic, biological potency of, 32; arsenical, 32; chlorinated hydrocarbon, 33–41, 65, 130, 163–7; organic phosphorus, 33–5, 41–5, 172, 174–7; systemic, 46–7; absorbed in plant tissues, 66–8; fatal to birds, 100–10, 113–20; in household use, 158–60; available to home gardeners, 160–2; storage in adipose tissue, 168–171; interaction between, 174–175; linked with mental disease, 175–7; research on, 225–6; modern, first medical use of, 232; bacterial, 249–52. *See also* Chemicals, Pesticides, and various chemicals by name

Insects, 'flareback' after spraying, 25, 220–24; disease-carrying, 26, 224; incidence of, under single-crop farming, 27; strains resistant to chemicals, 215; control of, 215; fecundity of, 215; held in check by natural forces, 217–219; parasitic, 218; population upsets caused by chemicals, 220–24; biological control of, 223, 226–8, 240–57; resistant to spraying, 229–37; agricultural, developing resistance of,

235; mechanism of resistance, 237–9; experiments with secretions of, as weapons, 246–8; male annihilation programmes, 248; ultrasonic sound as weapon against, 248–9; diseases of, as weapons against, 249–51; natural enemies as aid in control of, 252–7. *See also* various insects by name

IPC, 198, 207, 209

Iroquois County, Illinois, Japanese eradication programme in, 92–5

Irrigation waters, contamination of, 55–6

Jacob, F. H., 226

Japanese beetle, adverse side-effects of spraying, in Midwest, 90–6, 222–3; control of, in the eastern states, 96; milky disease of, 97–9, 250; total annual damage by, 223

Joachimsthal, lung cancer among workers at, 194

Joliet, Illinois, disastrous effects of heptachlor in, 92

Journal of Agricultural and Food Chemistry, 239

Kafue bream, 134

Klamath Lake, Lower and Upper, 55

Klamath weed, 84–5

Klinefelter's syndrome, 191

Knipling, Dr Edward, 241, 242, 245

Koebele, Albert, 252

Korea, lice develop resistance to DDT in, 233

Kuala Lumpur, Malaya, resistant mosquitoes at, 237

Kuboyama, 202

Lacewings, 218

Ladybugs, 218

Laird, Marshall, 224

Lawns, treated for crabgrass, 83–84

Lead, arsenate of, 66, 221, 227

Leaf roller, red-banded, 221

Leather Trades Review, 231

Lehman, Dr Arnold, 36, 38

Leukaemia, 206; chromosome abnormality in, 191; and pesticides as causative agents, 196, 199–203; rapid development of, 199; rising incidence of, 200, 206–7; DDT and case histories of, 200–1; in children, 207; as possible two-step process, 210

Levan, Albert, 205

Lice, body, as disease carriers, 231; resistance among, 232, 233–4

Life (Simpson, Pittendrigh, Tiffany), 187

Lime sulphur, resistance to, 229

Lindane, nitrification affected by, 64, 65; household use of, 159; effects on nervous system, 175; plant mutations caused by, 189; and blood disorders, 201, 202, 206

Liver, cellular damage caused by DDT, 36, 38; diseases of, caused by chlorinated naphthalenes, 39; function of, 171; effect of chlorinated hydrocarbons on, 171–2, 174, 208; role in sex hormone inactivation, 207; damage, and cancer development, 208–9

Long Island, effect of spraying for gypsy moth on, 144

Louisiana, fish mortality in, 131–132; reluctance to sign up for fire ant programme in, 156; sugarcane borer increased by fire and chemicals, 222

Lower Klamath Lake, California, 55

Lucky Dragon, tuna vessel, 202

McGill University, cancer research at, 208

Maine, brush spraying in, 74–5; forest spraying affects fish in, 127

Maine Department of Inland Fisheries and Game, 127

Malaria, flare-ups of, 235. *See also* Mosquitoes

Malathion, 44, 45, 171; symptoms of poisoning by, 161; effect on nervous system, 176

Malaya, resistance of mosquitoes in, 237–8

Male annihilation programmes, 249

Male sterilization technique, 241–245

Maleic hydrazide, 189

Malformations. *See* Defects, congenital

Mammals: killed by weeds sprayed with 2,4-D, 81; killed by aldrin, 91–2, 95; killed by dieldrin, 94; killed by insecticides in England, 118; killed by fire ant programme, 151–4; insecticides found in testes of, 184–5; effect of arsenic ingestion on, 196; cancer research on, 209. *See also* Antelope, Beaver, Cats, Coyotes, Deer, Moose

Mantis, praying, 217, 219

Marigolds, used for combating nematodes, 82

Marsh gas, 34

Matagorda Bay, insecticides threaten waters of, 136

Matthysse, J. G., 112

Max Planck Institute of Cell Physiology, 203

Mayo Clinic, lymph and blood diseases treated at, 200

Mealy bugs, 252

Mehner, John, 102, 104

Melander, A. L., 229

Melbourne, University of, 177

Melon fly, 244, 248

Mental disease, insecticides linked with, 175–7

Mental retardation, 191

Mesenteries, protective, 36

Metcalf, Robert, 215

Metchnikoff, Elie, 250

Methane, 34

Methoxychlor, 171, 175, 232

Methyl chloride, molecular structure, 34

Methyl-eugenol, 248

Michigan Audubon Society, 91

Michigan State University, robin population reduced by spraying at, 102–4

Microbial insecticides. *See* Bacterial insecticides

Migration, world-wide, of organisms, 27–8

Milk: human, insecticidal residues in, 37; pesticide residues in, 145–6, 153–4, 163

Milkfish, destroyed by spraying, 134–5

Milky disease, Japanese beetle, 97–9, 250

Miller, Howard C., 111

Mills, Dr Herbert R., 137-8
Minnesota, University of, 81
Miramichi River, 122; salmon affected by DDT spraying, 123-6
Mississippi Agricultural and Experiment Station, 156-7
Mites, soil, 62; spider, 220, 221; DDT spraying leads to increase of, in western forests, 221; in Nova Scotia, 226
Mitochondria, 179-80
Mitosis, 186
Mölln, Germany, forest programme in, 254
Mongolism, 190, 192
Montana, forest spraying in, 128
Montana Fish and Game Department, 128, 129
Moose, 72, 73
Mosquitoes, control of, and problem of fish conservation, 135; genetic effect of DDT on, 189; malaria-carrying, 224; as disease transmitters, 231; *Culex*, 232; resistant to DDT, 232, 234-5, 237; ultrasonic sound as weapon against, 248-9. *See also* Anopheles
Moth, Argentine, used in weed control, 86
Mothproofing, 160
Mount Johnson Island, 114-15
Mule deer, 71, 72
Muller, Dr Hermann J., 186, 188, 242
Müller, Paul, 35
Murphy, Robert Cushman, 100, 145
Mustard gas, 186
Mutagens, 49; chemical, 186, 189-92

Mutations, genetic, 185; caused by various chemicals, 188-9; caused by X-rays, 241. *See also* Genetic effect
My Wilderness: East to Katahdin (Douglas), 72

Naphthalenes, 39, 201
National Audubon Society, 101, 137
National Cancer Institute, 210. *See also* Hueper, Dr W. C.
Natural History Survey. *See* Illinois Natural History Survey
Nature, checks and balances of, 215-16
Nematode worms, marigolds used against, 82
Nervous system, effect of insecticides on, 172-7
New York State, Dutch elm disease control in, 110-12
New York Times, 160
Newsom, Dr L. D., 156
Nickell, Walter P., 89
Nicotine sulphate, 32, 227
Nissan Island, 224
Nitrification effect of herbicides on, 64-5
Nitrophenols, 201
Nova Scotia, biological control of orchard pests in, 226-8
Nuclear division. *See* Mitosis

Occupational Tumors (Hueper), 196
Oestrogens and cancer, 208, 209
Office of Vital Statistics, National, 149, 182, 183, 195, 200
Oklahoma Wildlife Conservation Department, 134
Oligospermia, crop dusters subject to, 185

Organic phosphates, 41–5; effects on nervous system, 172–177

Organisms, world-wide migration of, 27–8

Oxidation, cellular, 178–81; effect of insecticides upon, 181–5; and cancer research, 203–5

Oysters, 140

Pacific Flyway, 56

Pacific Science Congress, 134

Pallister, John C., 235

Paradichlorobenzene, 201

Paralysis, 'ginger', 176

Parathion, 42, 43–4, 45, 120, 160, 176

Pasteur, Louis, 194, 249

Patau, Dr Klaus, 191

Peanuts, insecticide-contaminated, 67

Pennsylvania, fish mortality in, 131

Penta (pentachlorophenol), 49, 181

Pest Control Institute, Springforbi, Denmark, 237

Pesticides, world-wide distribution of, 31–2; and blocking of process of oxidation, 182; as mutagens, 186, 188–9; as carcinogens, 196–203; indirect role in cancer, 209; and upset of insect populations, 220–24. *See also* Chemicals, Insecticides, and various chemicals by name

'Pheasant sickness', 119

Phenols: effect on metabolism, 181; genetic effects of, 189

Phillip, Captain Arthur, 85

Philippines, fish killed by spraying in, 135

Phosphates. *See* Organic phosphates

Phosphorylation, coupled, 181

Pickett, A. D., 226–7

Pittendrigh, Colin S., 187

Plankton, DDD accumulated by, 58; herbicides toxic to, 140

Plant killers. *See* Herbicides *and* Weed killers

Plants, importation of, 28

Pneumonia, chemical, 81

Poisoning, pesticide. *See* Disease, environmental

Poisons, availability of, to home-owners, 158–62

Poitevint, Dr Otis L., 153–4

Polistes wasp, 219

Pott, Sir Percival, 194

Price, Dr David, 168

Prickly pears, insect enemy used to control, 85–6

Prince Henry's Hospital, Melbourne, 177

Pyrethrins, 167

Pyrethrum, 32

Quail, 115, 152

Rabinowitch, Eugene, 179

Radiation, 24; as uncoupler, 181; and congenital deformity, 183; effect on living cell, 185; parallel between chemicals and, 185–7; and cancer, 193; sterilization of insects by, 241–5

Ragweed, 83

Ragwort, sprayed, attractive to livestock, 80

Rangelands, spraying of, 73

Ray, Dr Francis E., 195

'Reichenstein disease', 196

Reproduction: of birds, adversely affected by herbicides, 80; of birds, affected by DDT and related insecticides, 105, 115–117, 184–5; diminished, linked with interference with biological oxidation, 183

Reservoirs, insecticides in, 59

Residues, chemical, on food, 162–167

Resistance: of scale insects to lime sulphur, 229; of blue ticks to BHC, 230–31; of disease-carrying insects, 232; of houseflies to DDT, 232, 233; of various mosquitoes, 232, 234–5; of houseflies to BHC, 233; of body lice to DDT, 233–4; of malaria mosquitoes, 234–5; of ticks, 235; of German cockroaches, 236; of agricultural insects, 236; mechanism of, 237–9

Resurgence, insect, 25, 220–24

Rhoads, C. P., 208

Rhodesia, fish destruction in, 134

Rice fields, 119

Roadside spraying, 74–9

Robins: affected by spraying for Dutch elm disease, 101–6; reproduction affected by DDT, 116

Robson, William, 186

Rocky Mountain Arsenal, 53

Root borer, banana, 250–51

Rostand, Jean, quoted, 30

Rotenone, 167

Royal Society for the Protection of Birds, 117

Royal Victoria Hospital (McGill), cancer research at, 208

Rudd, Dr Robert, 115

Runner, G. A., 241

Ruppertshofen, Dr Heinz, 254, 255, 256

Rutstein, Dr David, 199

Ryania, 167, 227

Sagebrush, tragic consequences of campaign to destroy, 69–72, 76

St Johnswort. *See* Klamath weed

Salmon, Miramichi, affected by DDT spraying, 122–6; in British Columbia, killed by spraying, 129

San José scale, 229

Sardinia, insect resistance in, 233

Satterlee, Dr Henry S., 66

Sawflies, shrews as aid to control of, 255–6

Scale, San José, 229; cotton cushion, 223–4, 252

Schistosoma, 225

Schradan, 47

Schrader, Gerhard, 42

Schweitzer, Albert, quoted, 24

Screw-worms, eradicated through sterilization, 242–3

Seed treatment, effects of, in England, 117–19; in United States, 119

Sex hormones, imbalance of, and cancer development, 207–9

Sheldon, Illinois, effects of Japanese beetle eradication programme in, 92–6, 99

Shelf paper, insecticide-treated, 159

Shellfish, affected by chemicals, 140–41

Shepard, Paul, 28

Shrews, as aid in sawfly control, 255–6

Shrimp, 139–40

'Silo deaths', 81

Simpson, George Gaylord, 187

Single-crop farming, insect problems in, 27

Sloan-Kettering Institute, 205, 208

Snails, immune to insecticides, 224-5

Snow, John, 212

Soil, creation of, 61; organisms, 62-3; impact of pesticides on, 64-6; long persistence of insecticides in, 64-8

Soot, 32; as containing cancer-producing agent, 194

Sound, ultrasonic, as weapon against insects, 248-9

South-east Asia, mosquito control programmes threaten fish in, 134-5

Sparrow, house, relative immunity to some poisons, 151

Spider mites. *See* Mites

Spiders, as agents for biological control of insects, 254-5

Spraying, 'brush control', 73-6; selective, 78-9, 84; disastrous effect on wildlife, 87-8; aerial, 142-3; for gypsy moth, 144-7; modified, 227-8

Springforbi, Denmark, Pest Control Institute at, 237

Springtails, 62

Steinhaus, Dr Edward, 251-2

Sterility: caused by aldrin, 40; of grebes, 58; caused by insecticide poisoning, 105; of robins, 105; of eagles, 114-15; experimentally produced in birds, 190

Sterilization of male insects, as method of control, 241-5; by chemicals, 245

Strontium, 23, 206

Sugarcane borer, heptachlor increases damage by, 222

Super races, evolution of, 25

Swallows, 107

Swanson, Professor Carl P., 240

Sweeney, Joseph A., 110

Sweet potatoes, BHC-contaminated, 66-7

Syracuse, New York, Dutch elm disease in, 111

Syrphid fly, 217

Texas Game and Fish Commission, 135, 136

Ticks, developing resistance to chemicals, 230-31, 235

Tiffany, L. Hanford, 187

Tiphia vernalis, 97, 252

Tobacco, arsenic content of, 66

Tobacco hornworm, 248

Toledo, Ohio, Dutch elm disease in, 110

'Tolerances', 165-7

Toxaphene, toxic to fish, 51, 130, 131, 135; used against boll weevils, 132; and blood disorders, 202

Triorthocresyl phosphate, 176

Trout, liver cancer in, 210-11

Trouvelot, Leopold, 143

Tsetse fly, British experiments to eradicate, 244

Tule Lake, California, 55

Turkeys, wild, reduced by fire ant programme, 152

Turner, Neely, 29

Turner's syndrome, 191

2,4-D, spontaneous formation of, 54; nitrification interrupted by, 64; physiological effects, 80; curious effect on livestock, 80-81; nitrate contents of plants increased by, 80-1; as cause of unplanned changes in vegetation, 83; as uncoupler, 181-2;

plant mutations caused by, 189-90

2,4,5-T, 79

Typhus, DDT used against, 232; DDT ineffective against, 234

Ullyett, G. C., 228

Uncoupling, 181-2

U.S. Department of Agriculture: rulings on heptachlor, 67; Japanese beetle programme, 92, 93; research on milky disease, 99; campaign against fire ants, 147-54, 156-7; on moth-proofing, 160; estimates of Japanese beetle and corn borer damage, 223; on resistance of insects, 239; and development of male sterilization techniques, 241, 244-5

U.S. Fish and Wildlife Service: study of effects of DDT spraying, 52; reports on aldrin, 90; *Audubon Field Notes*, 101; concern over parathion, 120; study of budworm spraying, 128; study of fish with tumours, 210

U.S. Food and Drug Administration: regulations concerning chemical residues in food, 132, 163, 164, 165; pesticide residues in milk, 154; bans use of heptachlor on foods, 155; on dangers of chlordane, 159; jurisdiction, 165; recommendations on chemicals with cancer-producing tendencies, 197, 198

U.S. Forest Service, 72, 128, 220

U.S. Office of Plant Introduction, 28

United States Pharmacopeia, 176

U.S. Public Health Service, 54, 90, 130, 162-3

University of Melbourne, 177

University of Minnesota Medical School, 81

University of Wisconsin, 109; Agricultural Experiment Station, 81; research in chromosome abnormality, 191

Upper Klamath Lake, Oregon, 55

Urbana, Illinois, Dutch elm disease in, 110

Urethane, 189; as cancer-producing agent, 207, 210

Vedalia beetle, 223, 224, 252

Vegetation, roadside, spraying of, 73-6; importance of, 77-8; selective spraying of, 78-79

Viruses, as substitute for chemical insecticides, 251-2

Vitamins, protective role against cancer, 208-9

Wald, George, 178

Wallace, Dr George, 102, 103, 104, 105, 108, 115

Waller, Mrs Thomas, 145

Warblers, 107

Warburg, Professor Otto, 203-5

Wasp, *Tiphia vernalis*, 97, 252; muddauber, 217; horseguard, 217; *Polistes*, 219

Water: pollution by pesticides, 50-60; salt-shore, pesticidal pollution of, 137-41; polluted by detergents, 210. *See also* Fish

Waterford, Connecticut, trees injured by spraying at, 75

Waterfowl, spraying a threat to, 57-8, 138

Webworms, biological warfare against, 250, 251

Weed control, insect enemies used for, 84–6

Weed killers, 47–9. *See also* Crabgrass *and* Herbicides

Weevil, strawberry root, 67; boll, 132

West Virginia, bird population reduced in, 101

Wheeler Reservoir, Alabama, 133

Whiskey Stump Key, Florida, 137

Whitefish Bay, Wisconsin, decline of warblers in, 107

Wild cherry, sprayed, fatally attractive to livestock, 80

Wildlife losses from pesticides, 87–8; in Japanese beetle spraying, 91–2, 94–5; in Dutch elm disease spraying, 102–10; in England, 117–19; in rice fields, 119; in forest spray-ing, 123, 125, 126–30. *See also* Birds, Mammals, and various species

Winge, Ojvind, 206

Wisconsin, University of, 109; Agricultural Experiment Station, 81; chromosome research at, 191

Woodcocks, 106, 152

Woodticks, 235

World Health Organization, anti-malarial campaigns of, 40; Venezuelan cats killed by spraying of, 94; and problem of insect resistance, 230, 231

X-ray, sterilization of insects by, 241–5

Yellow fever, flare-ups of, 234–5

Yellowjackets, 217

Yellowstone River, fish destruction in, 127–8

CHERNOBYL PRAYER

Svetlana Alexievich

On 26 April 1986 the worst nuclear reactor accident in history occured in Chernobyl and contaminated as much as three quarters of Europe. While the official Soviet narrative downplayed the accident's impact, Svetlana Alexievich wanted to know how people understood it. She recorded hundreds of interviews with workers at the nuclear plant, refugees and resettlers, scientists and bureaucrats, crafting their monologues into a stunning oral history of the nuclear disaster. What their stories reveal is the fear, anger and uncertainty with which they still live but also a dark humour and desire to see the beauty of everyday life, including that of Chernobyl's new landscape. A chronicle of the past and a warning for our nuclear future, *Chernobyl Prayer* is a haunting masterpiece.

'A searing mix of eloquence and wordlessness ... From her interviewees' monologues she creates history that the reader, at whatever distance from the events, can actually touch' Julian Evans, *The Telegraph*

ONE-WAY STREET AND OTHER WRITINGS

Walter Benjamin

Walter Benjamin – philosopher, essayist, literary and cultural theorist – was one of the most original writers and thinkers of the twentieth century. This new selection brings together Benjamin's major works, including 'One-Way Street', his dreamlike, aphoristic observations of urban life in Weimar Germany; 'Unpacking My Library', a delightful meditation on book-collecting; the confessional 'Hashish in Marseille'; and 'The Work of Art in the Age of Mechanical Reproduction', his seminal essay on how technology changes the way we appreciate art. Also including writings on subjects ranging from Proust to Kafka, violence to surrealism, this is the essential volume on one of the most prescient critical voices of the modern age.

'There has been no more original, no more serious, critic and reader in our time' George Steiner

THE ROAD TO WIGAN PIER

George Orwell

'We are mistaken when we say that "It isn't the same for them as it would be for us", and that people bred in the slums can imagine nothing but the slums'

A searing account of George Orwell's observations of working-class life in the bleak industrial heartlands of Yorkshire and Lancashire in the 1930s, *The Road to Wigan Pier* is a brilliant and bitter polemic that has lost none of its political impact over time. His graphically unforgettable descriptions of social injustice, cramped slum housing, dangerous mining conditions, squalor, hunger and growing unemployment are written with unblinking honesty, fury and great humanity. It crystallized the ideas that would be found in Orwell's later works and novels, and remains a powerful portrait of poverty, injustice and class divisions in Britain.

'True genius . . . all his anger and frustration found their first proper means of expression in *Wigan Pier*' Peter Ackroyd, *The Times*

SOUTH

Ernest Shackleton

'We longed keenly for the day when we could begin this march, the last great adventure in the history of South Polar exploration'

In 1914 a party led by veteran explorer Sir Ernest Shackleton set out to make the first crossing of the entire Antarctic continent via the South Pole. But their initial optimism was short-lived as ice floes closed around their ship, *Endurance*, gradually crushing her to death and marooning the twenty-eight men on the polar ice.

Alone in the world's most unforgiving environment, Shackleton and his team began a brutal quest for survival. As the story of their journey across treacherous seas and a wilderness of glaciers and snowfields unfolds, the scale of their courage and heroism becomes movingly clear.

'One of the most harrowing survival stories of all time' Sebastian Junger

BEYOND THE PLEASURE PRINCIPLE

Sigmund Freud

In Freud's view we are driven by the desire for pleasure as well as by the desire to avoid pain. But the pursuit of pleasure has never been a simple thing. Pleasure can be a form of fear, a form of memory and a way of avoiding reality. Above all, as these essays show with remarkable eloquence, pleasure is a way in which we repeat ourselves.

The essays collected in this volume explore, in Freud's uniquely subtle and accessible style, the puzzles of pleasure and morality – the enigmas of human development.

'Freud's great legacy . . . brilliantly exposes the state of the psyche'
Mark Edmundson

IN AMERICA

Susan Sontag

The story of *In America* is inspired by the emigration to America in 1876 of Helena Modrzejewska, Poland's most celebrated actress, accompanied by her husband, Count Karol Chlapowski, her fifteen-year-old son, Rudolf, the young journalist and future author of *Quo Vadis*, Henryk Sienkiewicz, and a few friends; their brief sojourn in Anaheim, California; and Modrzejewska's subsequent triumphant career on the American stage under the name Helena Modjeska.

'A tour de force. . . . A magical accomplishment by an alchemist of ideas and words, images and truth' Michael Pakenham, *The Baltimore Sun*

THE FEMININE MYSTIQUE

Betty Friedan

When Betty Friedan produced *The Feminine Mystique* in 1963, she could not have realized how the discovery and debate of her contemporaries' general malaise would shake up society. Victims of a false belief system, these women were following strict social convention by loyally conforming to the pretty image of the magazines, and found themselves forced to seek meaning in their lives only through a family and a home. Friedan's controversial book about these women – and every woman – would ultimately set Second Wave feminism in motion and begin the battle for equality.

This groundbreaking and life-changing work remains just as powerful, important and true as it was forty-five years ago, and is essential reading both as a historical document and as a study of women living in a man's world.

'One of the most influential nonfiction books of the twentieth century' *The New York Times*

READING LOLITA IN TEHRAN

Azar Nafisi

Every Thursday morning in a living room in Iran, over tea and pastries, eight women meet in secret to discuss forbidden works of Western literature. As they lose themselves in the worlds of *Lolita*, *The Great Gatsby* and *Pride and Prejudice*, gradually they come to share their own stories, dreams and hopes with each other, and, for a few hours, taste freedom. Azar Nafisi's bestselling memoir is a moving, passionate testament to the transforming power of books, the magic of words and the search for beauty in life's darkest moments.

'I was enthralled and moved' Susan Sontag

OUT OF AFRICA

Karen Blixen

In 1914 Karen Blixen arrived in Kenya with her husband to run a coffee farm. Instantly drawn to the land, she spent her happiest years there until the plantation failed. Karen Blixen was forced to return to Denmark in 1931 and it was there that she wrote this classic account of her experiences. A poignant farewell to her beloved farm, *Out of Africa* describes her strong friendships with the people of her area, her affection for the landscape and animals, and great love for the adventurer Denys Finch-Hatton.

Written with astonishing clarity and an unsentimental intelligence, *Out of Africa* portrays a way of life that has disappeared for ever.

'Compelling . . . a story of passion . . . and a movingly poetic tribute to a lost land' *The Times*

THE ROAD TO SAN GIOVANNI

Italo Calvino

The Road to San Giovanni contains five autobiographical essays – fascinating expeditions through the memories of one of the greatest writers of the twentieth century. In these elegant meditations Calvino delves into his past, remembering awkward childhood walks with his father, a lifelong obsession with the cinema, and fighting in the Italian Resistance against the Fascists. He also muses on the social contract, language and sensations associated with emptying the kitchen rubbish and the shape he would, if asked, consider the world. These reflections on the nature of memory itself are engaging, witty and lit through with Calvino's alchemical brilliance.

'Brimming with Calvino's beautifully crafted prose, dry humour and continual questioning of his own writing and memory' *Observer*

EAST OF EDEN

John Steinbeck

'A man, after he has brushed off the dust and chips of his life, will
have left only the hard, clean questions: Was it good or was it evil?
Have I done well – or ill?'

'There is only one book to a man' Steinbeck wrote of *East of Eden*.
Set in the rich farmland of the Salinas Valley, California, this powerful, often brutal, novel follows the intertwined destinies of two
families – the Trasks and the Hamiltons – whose generations hope-
lessly re-enact the fall of Adam and Eve and the poisonous rivalry of
Cain and Abel. Here Steinbeck created some of his most memorable
characters and explored his most enduring themes: the mystery of
indentity; the inexplicability of love, and the murderous consequences of love's absence.

'A fantasia of history and myth . . . a strange and original work of art'
The New York Times Book Review

THE PLAGUE

Albert Camus

'This empty town, white with dust, saturated with sea smells, loud with the howl of the wind'

The townspeople of Oran are in the grip of a deadly plague, which condemns its victims to a swift and horrifying death. Fear, isolation and claustrophobia follow as they are forced into quarantine. Each person responds in their own way to the lethal disease: some resign themselves to fate, some seek blame, and a few, like Dr Rieux, resist the terror.

An immediate triumph when it was published in 1947, The Plague is in part an allegory of France's suffering under the Nazi occupation, and a story of bravery and determination against the precariousness of human existence.

'Enduring fiction has the power to grow into new kinds of timeliness' Boyd Tonkin, *Independent*

THE PLAGUE

Albert Camus